Frontiers in Information Systems

(Volume 1)

GIS: An Overview of Applications

Edited by

Ana Clàudia Teodoro

Department of Geosciences, Environment and Land Planning, Faculty of Science,
University of Porto, Rua Campo Alegre 4169-007 Porto, Portugal

Frontiers in Information Systems

Volume # 1

GIS: An Overview of Applications

Editor: Ana Cláudia Teodoro

ISSN (Online): 2589-3793

ISSN (Print): 2589-3785

ISBN (Online): 978-1-68108-611-8

ISBN (Print): 978-1-68108-612-5

Published by Bentham Science Publishers – Sharjah, UAE. All Rights Reserved.

General:

1. Any dispute or claim arising out of or in connection with this License Agreement or the Work (including non-contractual disputes or claims) will be governed by and construed in accordance with the laws of the U.A.E. as applied in the Emirate of Dubai. Each party agrees that the courts of the Emirate of Dubai shall have exclusive jurisdiction to settle any dispute or claim arising out of or in connection with this License Agreement or the Work (including non-contractual disputes or claims).
2. Your rights under this License Agreement will automatically terminate without notice and without the need for a court order if at any point you breach any terms of this License Agreement. In no event will any delay or failure by Bentham Science Publishers in enforcing your compliance with this License Agreement constitute a waiver of any of its rights.
3. You acknowledge that you have read this License Agreement, and agree to be bound by its terms and conditions. To the extent that any other terms and conditions presented on any website of Bentham Science Publishers conflict with, or are inconsistent with, the terms and conditions set out in this License Agreement, you acknowledge that the terms and conditions set out in this License Agreement shall prevail.

Bentham Science Publishers Ltd.
Executive Suite Y - 2
PO Box 7917, Saif Zone
Sharjah, U.A.E.
Email: subscriptions@benthamscience.org

**BENTHAM
SCIENCE**

CONTENTS

FOREWORD

Geographic Information Systems (GIS) environment permits the collection, analysis and integration of spatial data coming from different surveys. It is our pleasure to present this volume, where we can find several chapters about the applications of GIS to different Sciences.

Chapters in this book are related to specific topics, including:

1. Landscape planning
2. Tourism
3. Geological resources exploration
4. Environment
5. Sustainable planning
6. Biology
7. Human geography

This book aims at constituting a milestone and a bridge for the development and challenges for a better future.

This volume testifies the evolution of GIS applications during the last years, and summarizes the recent results of application to different domains which are very useful to the Society. The contributions of authors have explored several questions ranging from scientific to economic aspects, from professional applications to ethical issues, which all have a possible impact on society and territory.

We hope that you will be able to find stimulating contributions, which will support your research or professional activities.

The successful completion of this book was only possible because of the dedication and hard work of Ana Claudia.

Fernando Noronha
Department of Geosciences
Environment and Land Planning
University of Porto
Portugal

PREFACE

This book is the result of a compilation of several advances in the different applications of Geographic Information Systems (GIS) technology to distinct areas. GIS is, in its essence, an applied science. The applications presented in this book were selected in an attempt to cover some of the most representative areas of action. Of course, from the reader's perspective, some of them may be missed, but on the other hand, the reader may also be surprised by some of the applications covered by this book. From the start, the intention was to produce a compilation of GIS applications as heterogeneous as possible to show the universality of GIS. In this book, areas as different as land use and land cover changes, tourists' destinations, pegmatite mapping, forest management, spatial biology, environmental applications, health monitoring and also open source GIS tools are covered.

This book also intends to be an updated tool over the state-of-the-art of the different GIS intervention areas. In other words, readers will have access to the latest advances of GIS technology in different areas.

In my opinion, one of the major gaps in the present bibliography about GIS is the lack of a compilation of this type, *i.e.* bringing together several and distinct application areas rather than focusing only on one area.

I think that the target audience of this book could be very broad. First of all, the scientific community will find a compilation of the most recent advances in GIS science; also students who could use this book as an auxiliary textbook for their classes, and finally companies that develop GIS solutions.

This book also means to outline the new trends of GIS. Some of the applications described are quite well-known, but others are very different and very original. In these cases, the reader will gain a new perspective of GIS.

I hope that this book will inspire others to compile original GIS applications.

Many thanks to all my colleagues.

Ana Cláudia Teodoro
Department of Geosciences, Environment and Land Planning,
Faculty of Science, University of Porto,
Rua Campo Alegre 4169-007 Porto,
Portugal

List of Contributors

Alexandre Lima
Institute of Earth Science (ICT), Faculty of Sciences, University of Porto, Porto, Portugal

Ana Cláudia Teodoro
Department of Geosciences, Environment and Land Planning, Faculty of Sciences, University of Porto, Porto, Portugal
Earth Sciences Institute (ICT), Faculty of Sciences, University of Porto, Porto, Portugal

Cândida G. Vale
CIBIO Research Centre in Biodiversity and Genetic Resources, InBIO, Campus Agrário de Vairão, Rua Padre Armando Quintas, No 7. 4485-661 Vairão, , Universidade do Porto, Vila do Conde, Portugal

Carlos Cardoso Ferreira
Institute of Geography and Spatial Planning, Universidade de Lisboa, Lisboa, Portugal

Carlos Silva Neto
Institute of Geography and Spatial Planning, Universidade de Lisboa, Lisboa, Portugal

Charles Gumiaux
ISTO, UMR 7327, 1A rue de la Férollerie, Universitéd'Orléans, Orléans, France

Christian Evans
Physical Therapy, Midwestern University, Downers Grove IL, USA

David Silva
DGAOT, University of Porto, 687, R. Campo Alegre, Portugal

Eric Gloaguen
BRGM, ISTO, UMR 7327, av. Claude Guillemin, 45060 Orléans, France

Eusébio Joaquim Marques dos Reis
Institute of Geography and Spatial Planning, Universidade de Lisboa, Lisboa, Portugal

Fernando Noronha
Institute of Earth Science (ICT), Faculty of Sciences, University of Porto, Porto, Portugal

Francisco Reis Sacramento Gutierres
Eurecat, Centre Tecnològic de Catalunya, Big Data & Data Science Unit, Barcelona, Spain

Inês Boavida-Portugal
Department of Spatial Planning and Environment, Faculty of Spatial Sciences, University of Groningen, Groningen, The Netherlands

Jorge Rocha
Department of Spatial Planning and Environment, Faculty of Spatial Sciences, University of Groningen, Groningen, The Netherlands

José Alberto Álvares Pereira Gonçalves
Interdisciplinary Centre of Marine and Environmental Research (CIIMAR), University of Porto, Porto, Portugal

José Luís Zêzere
Institute of Geography and Spatial Planning, Universidade de Lisboa, Lisboa, Portugal

Juan Manuel Domingo Santos
Departamento de Ciencias Agroforestales, University of Huelva, E-21819 Palos de la Frontera, Spain

Lia Bárbara Cunha Barata Duarte
Department of Geosciences, Environment and Land Planning, Faculty of Sciences, University of Porto, Porto, Portugal
Earth Sciences Institute (ICT), Faculty of Sciences, University of Porto, Porto, Portugal

Neftalí Sillero
CICGE, Centro de Investigação em Ciências Geo-Espaciais, Faculdade de Ciências da, Universidade do Porto (FCUP), Vila Nova de Gaia, Portugal

Paulo Jorge Zuzarte de Mendonça Godinho-Ferreira
Strategic Unit for Research and Services of Agrarian and Forest Systems & Plant Health, National Institute for Agrarian and Veterinarian Research, Oeiras, Portugal

Rubén Fernández de Villarán San Juan
Departamento de Ciencias Agroforestales, University of Huelva, E-21819 Palos de la Frontera, Spain

Sarah Deveaud
BRGM, ISTO, UMR 7327, av. Claude Guillemin, 45060 Orléans, France

Sungsoon Hwang
Department of Geography, DePaul University, Chicago, IL, USA

Timothy Hanke
Physical Therapy, Midwestern University, Downers Grove IL, USA

Wouter Beukema
Department of Pathology, Bacteriology and Avian Diseases, Faculty of Veterinary Medicine, Ghent University, Salisburylaan 133, 9820, Merelbeke, Belgium

Predicting Land Use and Land Cover Changes for Landscape Planning: An Integration of Markov Chains and Cellular Automata Using GIS

Francisco Reis Sacramento Gutierres[1,*], **Paulo Jorge Zuzarte de Mendonça Godinho-Ferreira**[2], **Eusébio Joaquim Marques dos Reis**[1] and **Carlos Silva Neto**[1]

[1] *Eurecat, Centre Tecnològic de Catalunya, Big Data & Data Science Unit, Barcelona, Spain*

[2] *Strategic Unit for Research and Services of Agrarian and Forest Systems & Plant Health / National Institute for Agrarian and Veterinarian Research, Oeiras, Portugal*

Abstract: The spatial dynamics of landscapes are the consequence of a multiplicity of relations among physical, biological and social forces. So, it is essential the assessment of the driving forces related to Land Use and Land Cover changes (LULC) to understand the change process. The stochastic modeling technique in Geographical Information System (GIS) - Markov Chain (MC) analysis and Cellular Automata (CA) allows the predictions of future changes based on changes that have occurred in the past. This chapter aims to present a dynamic simulation model for LULC changes in 'Sado Estuary' and 'Comporta-Galé' Natura 2000 Sites (Portugal) for the beginning of the second half of the XXI century by using MC and CA. Regarding the quantification of the fragmentation processes and LULC changes in 'Sado Estuary' and 'Comporta-Galé' Natura 2000 Sites, these models are able to reveal non-obvious trends in the data and to describe ecological patterns. From an applied research point of view, this approach is useful to identifying adequate planning and management strategies for coastal ecosystems, for monitoring and planning natural and protected environments.

Keywords: Cellular Automata, GIS, Land Change Modeler, Land use and Land Cover, Markov Chains, Sado Estuary and Comporta Galé Natura 2000 Sites.

INTRODUCTION

The spatial dynamics of landscapes are the consequence of a multiplicity of relations among physical, biological and social forces [1, 2]. In recent years the Land Use and Land Cover (LULC) changes modeling are being viewed as crucial taking into account the anthropogenic impact [3, 4].

---

* **Corresponding author Francisco Reis Sacramento Gutierres:** Eurecat, Centre Tecnològic de Catalunya, Big Data & Data Science Unit, Barcelona, Spain; Tel/Fax: +34 677201569; E-mail: francisco.sacramento@eurecat.org

Ana Cláudia Teodoro (Ed.)

Spatial models due to their temporal dimension are fundamental to understand this process [5]. Therefore exists diverse methods for modeling spatial dynamics, and can be categorised in Stochastic modeling - Markov Chain (MC) and Cellular Automata (CA), Agent-Based, Neural Network and Fractal modeling and others, according to the complexity and nonlinearity [6 - 9].

Spatially explicit simulation models support the test hypotheses of the landscape evolution considering different states. Therefore, the importance of adopting a dynamic simulation model for landscape planning decisions and to design new land-management techniques has been widely recognized in the recent decades [2, 10 - 12].

The assessment of the landscape dynamics, and likewise for a sustainable landscape management, requires an evaluation of the landscape quality and quantity. In this sense, the Geographical Information System (GIS) makes available novel tools for this analysis and evaluation. Thus, the stochastic models MC – CA can easily be incorporated into GIS [13 - 15]. CA allows the simulation of static entities in spatial models by diffusion, and encodes spatial structures. GIS based modeling approach can be implemented to understand the LULC changes related to biotopes, their rate of change and dynamic patterns. The stochastic modeling technique in GIS (MC analysis and CA) can simulate and predict the LULC changes trends and characteristics based on events that have occurred in the past [2, 10, 16 - 20].

This chapter aims to present a dynamic simulation model for LULC changes in 'Sado Estuary' and 'Comporta-Galé' Sites (Portugal) for the beginning of the second half of the XXI century by using MCs and CA. The specific objectives of the study are, first, to elaborate the drivers and magnitude of the long term LULC changes in a very important area for nature conservation. Secondly, we seek to display the outcomes of the future LULC transformations. Finally, to provide recommendations for the sustainable landscape planning in Natura 2000 areas. The analytical steps are threefold: (1) understand the landscape patterns with special attention to the transitions in coastal habitats; (2) development of a stochastic modeling technique in GIS, MC Analysis and CA with categorized LULC data taken over a 17-year period (1990-2007) of three sectors and; (3) associate these aspects to landscape fragmentation process, management practices (*e.g.* deforestation and anthropic pressure on the coastal zones and damages caused by the pine tree nematode), socio-economic factors and climate changes.

Study Area

The study area comprises the sites of Natura 2000 Network 'Sado Estuary' and 'Comporta-Galé' (total area of 63 018 ha). According to the biogeographical typology of Rivas-Martínez [21] it belongs to the "Ribatagano Sadense" Sector and to the Lower thermomediterranean bioclimate. The flora and the vegetation of the study area have been arranged in six broad biogeosystems characterized by particular floristic communities and lito-morpho-pedological conditions: beaches and coastal dunes under the influence of salt spray and sea breezes; dunes and inland sandy coverings; conglomerate, gritty coastal cliffs; surfaces made up "Marateca formation" sandstone and conglomerate; peat-bogs; and marshy areas [22 - 25]. The analysis of the landscape dynamics will be centred on three sectors within the 'Sado Estuary' and 'Comporta-Galé' Sites of Community Importance (SIC) (Fig. **1**). Sector 1 (21 624 ha), located northeast of Sado Estuary, representative of 'Sado Estuary (93)', 'Pegões Sands (92)' and 'Sado Heathlands (94)' landscape units; sector 2 (36 225 ha), located at Comporta, representative of 'Sado Estuary (93)', and 'Pine belts of Alentejo Litoral (95)' landscape units; and sector 3 (37 296 ha), located at Santo André and Sancha Lagoons Natural Reserve and representative of 'Pine belts of Alentejo Litoral (95)' landscape unit [26].

DATA AND METHODOLOGY

Modeling the LULC changes, in three sectors within the 'Sado Estuary' and 'Comporta-Galé' SIC, was conducted in three phases. Firstly, the data collection and LULC layers preparation for several years covering the study area. In second, the LULC changes were investigated. In the final phase, the factors that affect LULC patterns were determined, and the LULC changes founded on past changes and the factors were simulated. The methodology followed in this work is presented in Fig. (**2**).

Data Sets

National land-register for 1958 ('Serviço de Reconhecimento e Ordenamento Agrário (SROA)'), 1990 and 2007 ('Direção-Geral do Território (DGT)'), with 1/25 000 scale, were used in this research. Spatial data were organized and processed in ArcGIS 10.2 (ESRI) [27] software as raster maps with 100 m resolution. The projection used was the European Terrestrial Reference System 1989 (PT-TM06/ETRS89).

Classification Scheme

The purpose of this study was to analyze the local trend of LULC types for the selected sectors. Therefore, the adopted LULC classification scheme encompassed fourteen detailed classes. The legend of the LULC types maps took into account the heterogeneity of sources (national land-register of 1958, 1990 and 2007) and the new Corine Land Cover (CLC) Level 5 Nomenclature for Portugal according to Guiomar *et al.* [28] and adapted to the study area (Table **1**).

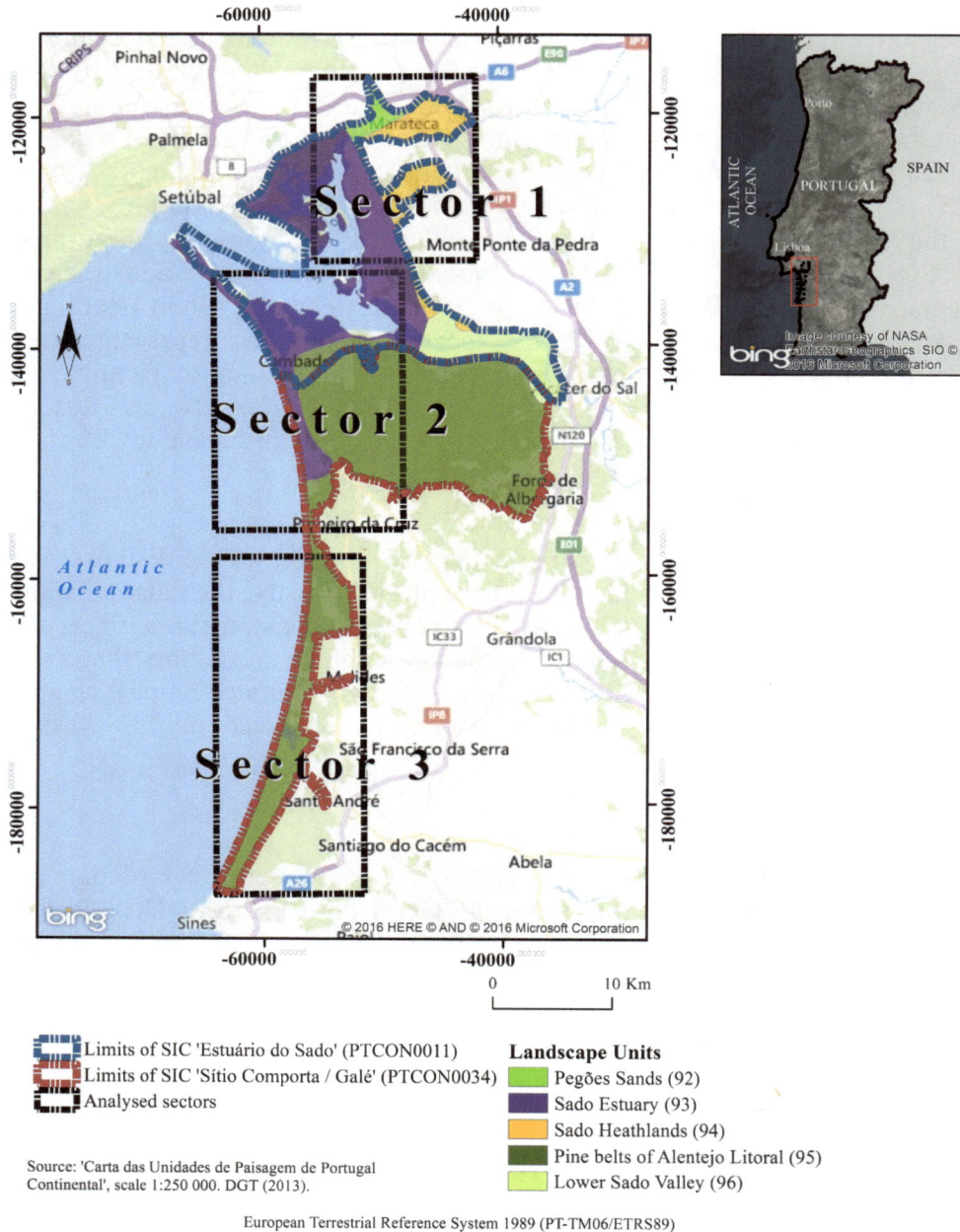

Fig. (1). Location map of study area ([26]).

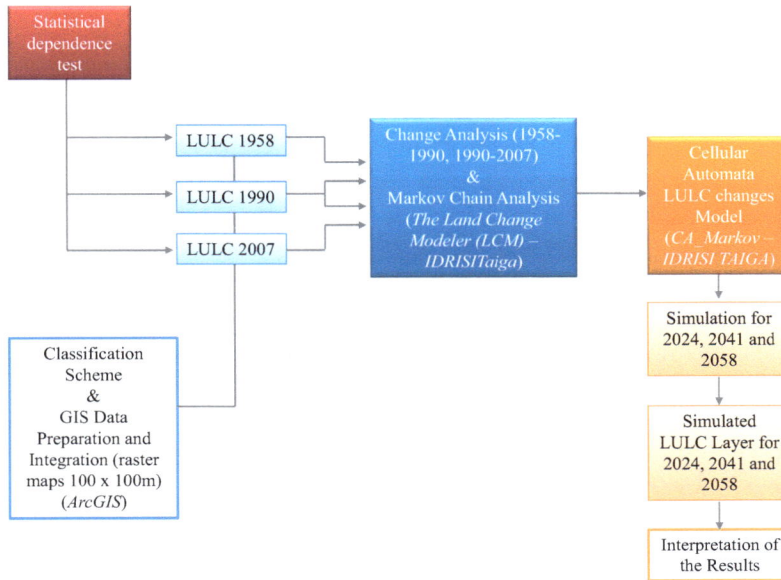

Fig. (2). Workflow for the simulation of LULC changes.

Table 1. Land Use and Land Cover classes.

Land use and land cover types (LULCT)	CLC (Level 2 and 3)	Description
Class 1	1.1, 1.2, 1.3, 1.4	Artificial areas
Class 2.1	2.1	Arable land
Class 2.2	2.2	Permanent crops
Class 2.3	2.3	Pastures
Class 2.4	2.4	Heterogeneous agricultural areas
Class 3.1	3.1.1	Broad-leaved forest
Class 3.2	3.1.2	Coniferous forest
Class 3.3	3.1.3	Mixed forest
Class 3.4	3.2	Shrub and/or herbaceous vegetation association
Class 3.5	3.3	Open spaces with little or no vegetation
Class 4.1	4.1	Inland wetlands
Class 4.2	4.2	Coastal wetlands
Class 5.1	5.1	Inland waters
Class 5.2	5.2	Marine waters

Stochastic Modeling Technique in GIS (MC Analysis and CA)

According to Gutierres [2] this GIS based modeling approach can be used to understand the LULC changes related to Natura 2000 habitats, their rate of change and dynamic patterns. This stochastic model allows the predictions of future change based on changes that have occurred in the past [10, 16 - 20, 29, 30]. On the basis of observed data over certain time periods MC analysis computes the probability that a cell will change from one LULC class (state) to another within a specified period of time. The probability of change from one state to another state is called a transition probability [31]. According to Hegde *et al.* [29], CA involves a simulation environment characterized by a raster structure, in which a set of transition rules define the attribute of each specified cell considering the attributes of cells in its neighbourhood cells. The state of a cell is defined by the preceding states of a surrounding neighbourhood of cells. Thus, CA adds spatial contiguity as well as knowledge of the expected spatial distribution of transitions to MC Analysis [29, 32, 33].

According Cabral *et al.* [34], a stochastic process produce sequences of random variables $\{X_n, n \in T\}$ by probabilistic laws. For instance, index n represents time. The process is described as discrete in time and $T = \{0, 5, 10\}$ years approximately, which can be looked as appropriate time unit for studying LULC changes. If the stochastic process is presented as a Markovian process, it implies that the sequence of random variables will be generated by the Markov property (1):

$$P [X_{n+1} = a_{in+1} \mid X_0 = a_{i0}, ..., X_{in} = a_{in}] = P [X_{in+1} = a_{in+1} \mid X_{in} = a_{in}] \qquad (1)$$

where the double index means for $n \in T$ and $T = \{0, 5, 10\}$ and i the range of possible values that a_i can assume, in this study the 14 classes defined formerly. Once the variety of possible values for a_i is either finite or infinite denumerable, the Markovian process may be mentioned as a MC. To prove that LULC changes in the three selected sectors is a Markovian process, one must demonstrate that: there is a statistical dependence between X_{n+1} and X_n (2); and that statistical dependence is a first-order Markov process (3).

$$P (X_n = a_n \mid X_{n-1} = a_{n-1}) \neq P (X_n = a_n) \times P (X_{n-1} = a_{n-1}) \qquad (2)$$

$$P (X_n = a_n \mid X_{n-1} = a_{n-1}) = P (X_n = a_n, X_{n-1} = a_{n-1}) / P (X_{n-1} = a_{n-1}) \qquad (3)$$

where X is a random variable and n represents the time.

The main hypothesis to be tested, H0: the null hypothesis is that the LULC in the analysed sectors is generated by a first order Markov process (outlines the transition from a class to any other does not involve intermediate transitions to other states).

To prove H0 two secondary hypotheses must be confirmed: H1 - LULC in distinct time periods is not statistically independent and H2- LUCC in the study area is a Markov process.

The statistical dependence can be verified as in a contingency table [2, 35] revealing the LULC changes between X_n and X_{n-1}. In this study, this test refers to the LULC changes occurred between 1990 and 2007. To deduce from the association or independence among the LULC classes in different years from the contingency table, the random variable, with the chi-square distribution will be expressed by Equation (4):

$$X^2 = \sum_i \sum_j ((N_{ij} - M_{ij})^2 / M_{ij}) \qquad (4)$$

where N will be the contingency matrix presenting the LULC changes between 1990 and 2007, and M the contingency matrix with the probable values of change considering the independence hypotheses [35]. X^2 measures the distance among the observed values of LULC changes and the probable ones assuming independence and must be sufficient high to prove (Equation 2), for 169 degrees of freedom. The identical non-parametric test was applied to test the Markov property. In this case, the values to be compared with the observed ones were calculated from the Chapman-Kolmogorov (Equation 5) [30], considering that these variables are produced by a first-order Markov process:

$$P(X_n = a_n \mid X_m = a_m) = P(X_1 = a_1 \mid X_m = a_m) \times P(X_n = a_n \mid X_1 = a_1), 0 \le m \le 1 \qquad (5)$$

In respect to this study, the Chapman-Kolmogorov equation assumes that transition probabilities from years 1958 to 2007 can be determined by multiplying the transition probabilities matrix from years 1958 to 1990 by the transition probabilities matrix from years 1990 to 2007.

Thus, were developed a model of LULC changes, through the integration of MC Analysis and CA using GIS, with categorized LULC data derived from the National land-register taken over a 17-year period (1990-2007) of each sector. The CA simulates the LULCT predictions by inferring from the past and present states taking into account time-series imagery and the transition probability matrix 1990-2007, which allows the landscape composition and configuration forecasting to the years 2024, 2041 and 2058.

To predict the LULC changes in the three sectors, the 'The Land Change Modeler (LCM)' and 'IDRISI CA_Markov' modules of IDRISI TAIGA [36] were used.

RESULTS

Change Analysis and Markov Chain Analysis - Hypothesis Testing

As mentioned in the methodology section, the main hypothesis to be verified is that LULC changes in the study area is generated by a first order Markov process. LULC maps were created, respectively, for years 1958, 1990 and 2007 for the three sectors considered (Figs. **3**, **4** and **5**). For the purpose of this analysis, three contingency tables were considered to quantify LULC changes between years 1958 and 1990, 1958 and 2007 and 1990 and 2007 for each sector.

Broad-leaved forest (Cl. 3.1) (from 10957 to 5979 ha), arable land (CL. 2.1) (from 2489 to 1812 ha) and coastal wetlands (Cl. 4.2) (from 2285 to 1481 ha) have shown a consistent decrease between 1958 and 1990. As shown from the maps of Fig. (**3**), there has been an increase of mixed forest (Cl. 3.3) (from 1759 to 6081 ha) and heterogeneous agricultural areas (Cl. 2.4) (401 to 1593 ha). Marine waters (Cl. 5.2) (1163 to 1928 ha) have also shown a consistent increase between this period. The broad-leaved forest (Cl. 3.1) lost area until 1990 with a decrease rate of about 45%, standing out the losses occurred to mixed forest (Cl. 3.3), with a gain of 4366 ha. It should also be pointed out the conversion of 933 ha of coastal wetlands (Cl. 4.2), due to the continued erosion of the salt marsh habitats, to marine waters (Cl. 5.2).

Between 1990 and 2007 there has been an increase of heterogeneous agricultural areas (Cl. 2.4) (from 1593 to 5371 ha) and a decrease of forests classes (Cl. 3.1 and Cl. 3.3) (5979 to 680 ha and 6081 to 5970 ha, respectively). Because of the successive decrease of these classes, shrub and/or herbaceous vegetation association (Cl. 3.4) have dynamically increased, from 377 to 1870 ha, in this study period. Regarding the earlier period, the decrease in coastal wetlands (Cl. 4.2) (from 1481 to 1052 ha), related with the retreat of salt marshes, is the result of transactions occurred mostly in marine waters (Cl. 5.2) (gained 596 ha), however the coastal wetlands (Cl. 4.2) increased 45 ha due to arable land (Cl. 2.1). This is indicative of encroachment of salt marsh areas towards the agricultural lands.

LULC changes analysis in sector 2 shows the remarkable increase in coniferous forest (CL. 3.2) (from 8689 to 11667 ha), the decrease of arable land (Cl. 2.1) (from 3929 to 2174 ha) and of open spaces with little or no vegetation (Cl. 3.5) (from 2921 to 1022 ha) between 1958 and 1990 (Fig. **4**).

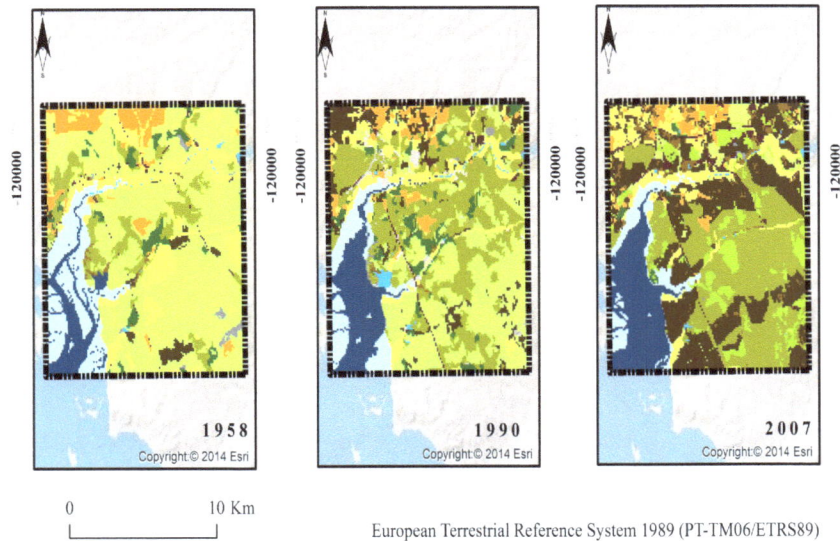

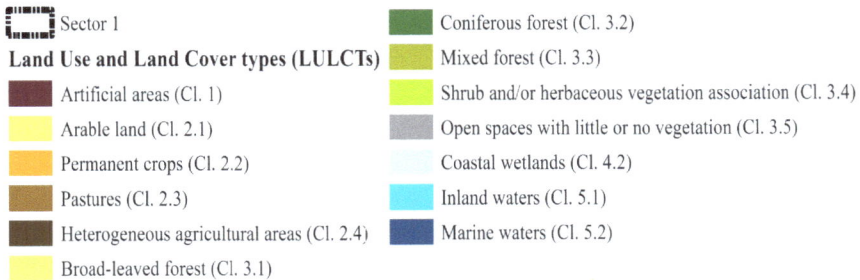

European Terrestrial Reference System 1989 (PT-TM06/ETRS89)

Sector 1

Land Use and Land Cover types (LULCTs)

Artificial areas (Cl. 1)

Arable land (Cl. 2.1)

Permanent crops (Cl. 2.2)

Pastures (Cl. 2.3)

Heterogeneous agricultural areas (Cl. 2.4)

Broad-leaved forest (Cl. 3.1)

Coniferous forest (Cl. 3.2)

Mixed forest (Cl. 3.3)

Shrub and/or herbaceous vegetation association (Cl. 3.4)

Open spaces with little or no vegetation (Cl. 3.5)

Coastal wetlands (Cl. 4.2)

Inland waters (Cl. 5.1)

Marine waters (Cl. 5.2)

Source:
'Carta Agrícola e Florestal de Portugal do Serviço de Reconhecimento e Ordenamento Agrário', scale 1:25 000.
SROA (1958).

'Carta de Uso e Ocupação do Solo de Portugal Continental (COS'90 e COS'07 - Nível 2)', scale 1:25 000. DGT (1990, 2007).

Fig. (3). LULC maps for year 1958, 1990 and 2007 for sector 1.

According to the results, the coastal wetlands (Cl. 4.2) area decreased from 3079 ha in the year 1958 to about 2448 ha in 1990, whereas during the same period, the mixed forest (Cl. 3.3), shrub and/or herbaceous vegetation association (Cl. 3.4) and marine waters (Cl. 5.2) increased from 304 to 615 ha, 0 to 732 ha and 14833 to 15478 ha, respectively. The null value of shrubland class (Cl. 3.4) in 1958 is explained due to its absence in the Agroforestry map's legend ('SROA, 1958'), however, it should be stressed that this occupation existed and would be associated with woodlands.

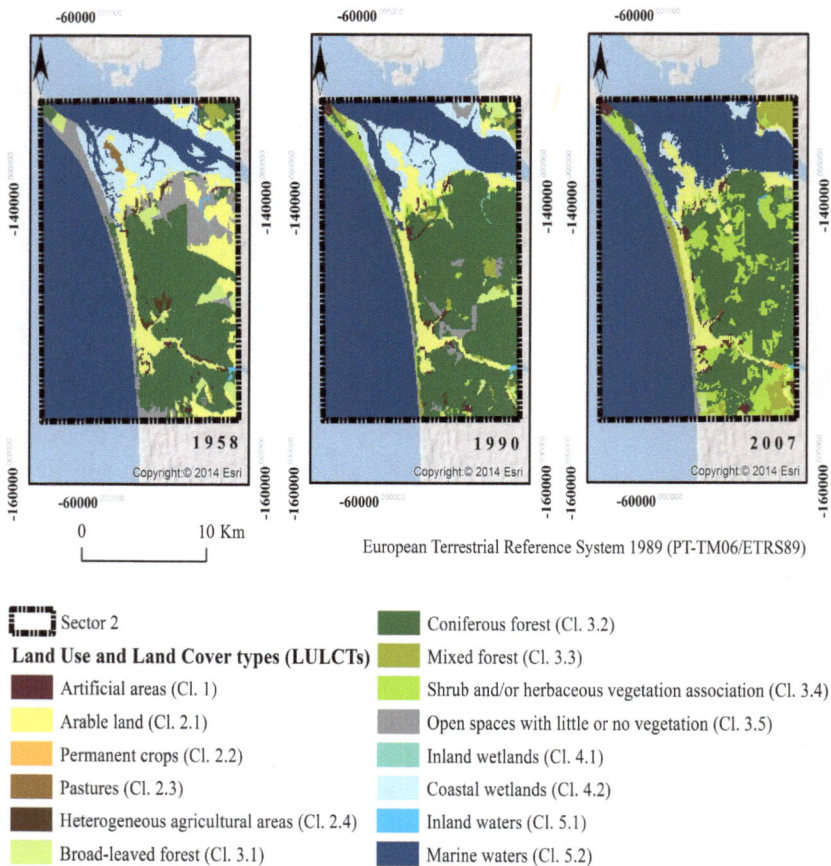

Source:
'Carta Agrícola e Florestal de Portugal do Serviço de Reconhecimento e Ordenamento Agrário', scale 1:25 000. SROA (1958).

'Carta de Uso e Ocupação do Solo de Portugal Continental (COS'90 e COS'07 - Nível 2)', scale 1:25 000. DGT (1990, 2007).

Fig. (4). LULC maps for year 1958, 1990 and 2007 for sector 2.

European Terrestrial Reference System 1989 (PT-TM06/ETRS89)

Sector 3	Coniferous forest (Cl. 3.2)
Land Use and Land Cover types (LULCTs)	Mixed forest (Cl. 3.3)
Artificial areas (Cl. 1)	Shrub and/or herbaceous vegetation association (Cl. 3.4)
Arable land (Cl. 2.1)	Open spaces with little or no vegetation (Cl. 3.5)
Permanent crops (Cl. 2.2)	Inland wetlands (Cl. 4.1)
Pastures (Cl. 2.3)	Coastal wetlands (Cl. 4.2)
Heterogeneous agricultural areas (Cl. 2.4)	Inland waters (Cl. 5.1)
Broad-leaved forest (Cl. 3.1)	Marine waters (Cl. 5.2)

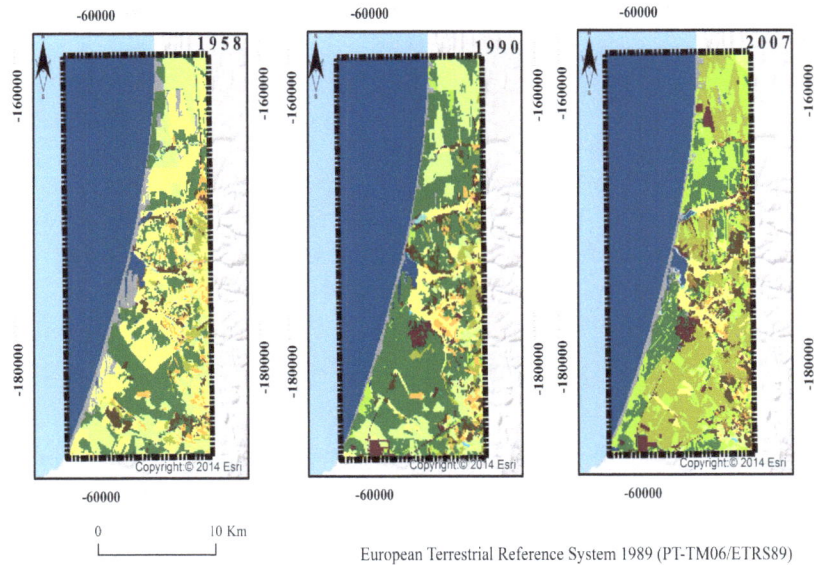

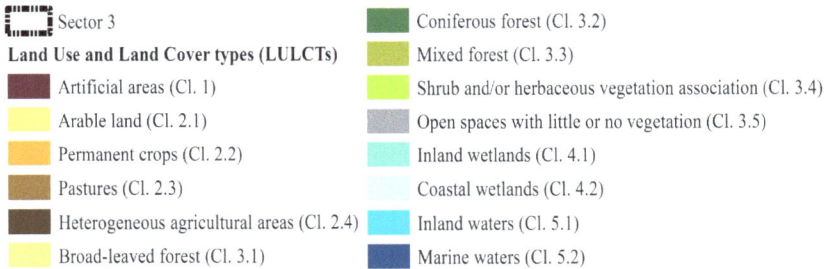

Source:
'Carta Agrícola e Florestal de Portugal do Serviço de Reconhecimento e Ordenamento Agrário', scale 1:25 000. SROA (1958).

'Carta de Uso e Ocupação do Solo de Portugal Continental (COS'90 e COS'07 - Nível 2)', scale 1:25 000. DGT (1990, 2007).

Fig. (5). LULC maps for year 1958, 1990 and 2007 for sector 3.

Coniferous forest (CL. 3.2) observed in 1990 (11667 ha) resulted from the conversion of 1882 ha of arable land area (Cl. 2.1), 1704 ha of open spaces with little or no vegetation (Cl. 3.5) and 244 ha of broad-leaved forest (Cl. 3.1).

It should also enhance the conversion of a vast area of open spaces with little or no vegetation (Cl. 3.5) (531 ha) to shrub and/or herbaceous vegetation association (Cl. 3.4).

Similar to sector 1, the coastal wetlands (Cl. 4.2) observed in 1958 (with 3079 ha) decreased until 1990 (with 2448 ha), whereas the losses occurred to the marine waters (Cl. 5.2) (gained 821 ha).

The spatial extent of coniferous forest (CL. 3.2) significantly decreased between 1990 until 2007 (from 11667 to 8530 ha) but the shrublands (Cl 3.4) (gained 3000 ha) and mixed forest (Cl. 3.3) (gained 901 ha) increased almost in the same extent reversely. The results reflect the anthropic pressures, namely the pine wood nematode and construction of tourist complexes, on wooded dunes with *Pinus pinea* and/or *Pinus pinaster* subsp. *atlantica* forests (Natura 2000 habitat '2270*'), and the consequent expansion of xerophytic shrublands.

Although this loss area in coniferous forest (CL. 3.2) until 2007 was evident, there were some gains in this class due to transition areas of broad-leaved forest (Cl. 3.1) (384 ha) and open spaces with little or no vegetation (Cl. 3.5) (383 ha).

During this period, mixed forests (Cl. 3.3) increased from 615 to 1380 ha, while the arable land (Cl. 2.1) slightly reduced their footprint in the landscape (from 2174 to 2013 ha).

Coastal wetlands (Cl. 4.2) also lost area (stronger than in the previous period) as a result of transitions to marine waters (Cl. 5.2) (from 2448 to 675 ha).

The sector 3 presented different aspects of LULC changes (Fig. **5**) between 1958 and 1990. According to the results, the arable land (Cl. 2.1) decreased from 6901 ha in 1958 to about 1998 ha in 1990. The same phenomenon occurred in open spaces with little or no vegetation (Cl. 3.5), with a reduction from 1510 to 602 ha. Furthermore, the surface area of heterogeneous agricultural areas (Cl. 2.4) (from 981 to 1375 ha), broad-leaved forest (Cl. 3.1) (from 3882 to 4043 ha), coniferous forest (Cl. 3.2) (from 6212 to 10307 ha) and mixed forest (Cl. 3.3) (from 960 to 1155 ha) varied greatly in referred period.

Arable land (Cl. 2.1) observed in 1990 resulted mainly from the conversion of 3139 ha to coniferous forest (Cl. 3.2), 784 ha to broad-leaved forest (Cl. 3.1), 644 ha to heterogeneous agricultural areas (Cl. 2.4) and 274 ha to mixed forest (Cl. 3.3).

As shown from the maps of Fig. (**5**), there has been an increase of heterogeneous agricultural areas (Cl. 2.4) due to transition areas of arable land (Cl. 2.1) (644 ha) and permanent crops (Cl. 2.2) (226 ha). The main changes observed for the time period of 1958 to 1990 was the decrease of coniferous forest (Cl. 3.2) (lost 256 ha) and arable land (Cl. 2.1) (lost 212 ha) due to urbanization from 156 to 694 ha.

Open spaces with little or no vegetation (Cl. 3.5) lost area until 1990 (from 1510 to 602 ha), standing out the losses occurred to coniferous forest (Cl. 3.2) (gained 689 ha), and to shrub and/or herbaceous vegetation association (Cl. 3.4) (gained 224 ha).

LULC changes analysis in this sector between 1990 and 2007 shows the remarkable decrease in coniferous forest (CL. 3.2) (from 10307 to 3390 ha) and broad-leaved forest (Cl. 3.1) (from 4043 to 591 ha); and the increase of mixed forest (Cl. 3.3) (from 1155 to 6064 ha) and shrub and/or herbaceous vegetation association (Cl. 3.4) (from 575 to 5522 ha). This LULC change pattern is explained by the dissemination of the pine wood nematode and the consequent expansion of xerophytic shrublands.

It should be noted that in the agricultural classes (Cl. 2.1, 2.2, 2.3 and 2.4) occurred transitions between them, such as the loss of 210 ha of arable land (Cl. 2.1) to heterogeneous agricultural areas (Cl. 2.4); as well as for shrubland class (Cl. 3.4), which gained 232 ha from Cl. 2.4, and for forest classes (Cl. 3.1, Cl. 3.2 and Cl. 3.3), where the mixed forest gained 131 ha from the heterogeneous agricultural areas (Cl. 2.4).

The value obtained to quantity the association among the contingency table 1958-2007 inside each sector and the Chapman-Kolmogrov equation was 25.96 (sector 1), 32.39 (sector 2) and 19.53 (sector 3). These values are evidently underneath the critical value of the distribution for a significance level of 0.950 which is 150.14. This outcome confirms the assumption that LULC change is a Markovian process inside the analyzed sectors.

Cellular Automata LULC Estimates for Years 2024, 2041 and 2058

With the 1990 and 2007 layers were determined the areas of LULC classes, which allowed the definition of the initial vector (1990 LULC composition), the transition matrix for the 1990/2007 period and the corresponding forecast equilibrium vectors (LULC composition for 2007, 2024, 2041 and 2058).

The stochastic model was implemented to estimate LULC change dynamics for years 2024, 2041 and 2058. This model consider that the transitional conditional probabilities between 1990/2007 are correspondent to the transitional conditional probabilities between 2007/2024, 2024/2041 and 2041/2058 and it spatialization is completely random.

In the studied sectors the predictions (2024, 2041 and 2058), derived from CA simulation (CA_Markov), indicate that some stability in the LULC classes will be attained in 2058 (Figs. **6**, **7** and **8**).

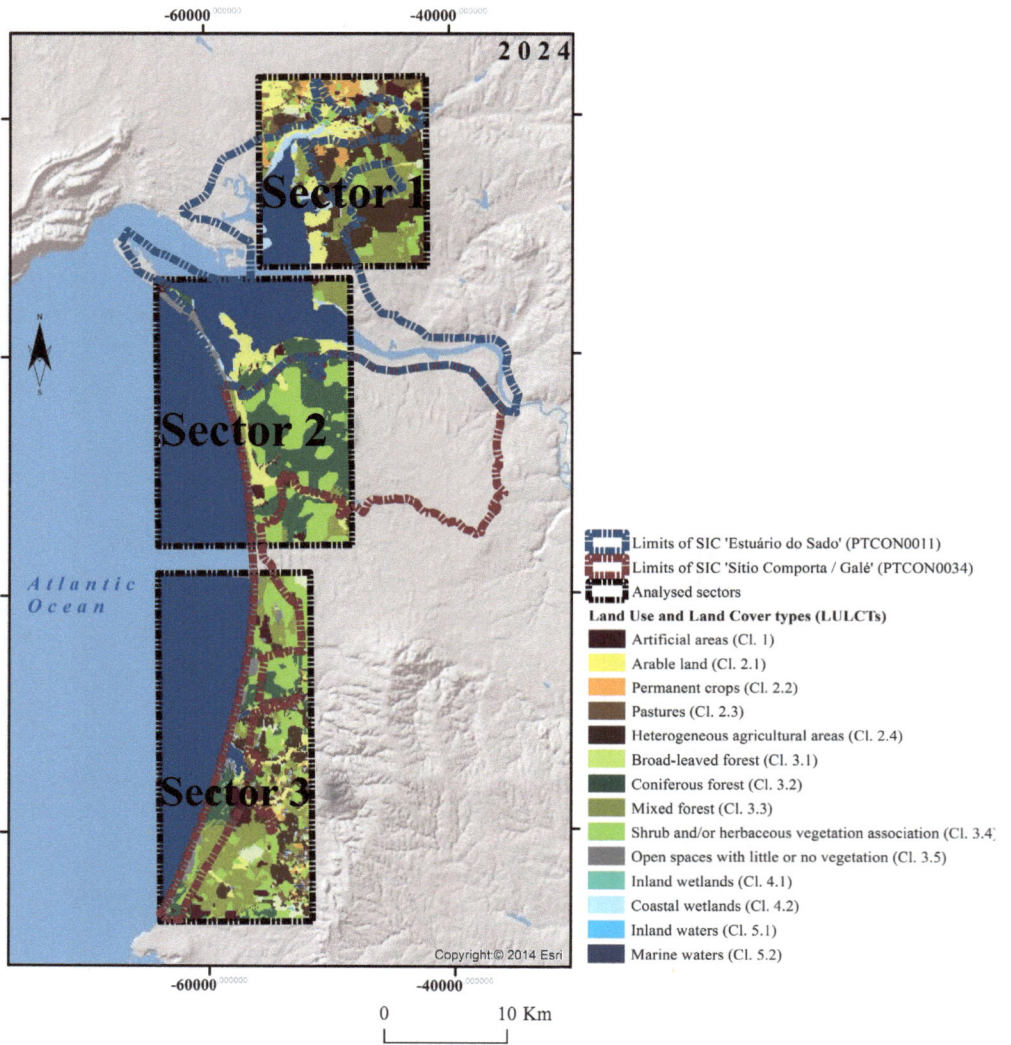

European Terrestrial Reference System 1989 (PT-TM06/ETRS89)

Source:
'Carta Agrícola e Florestal de Portugal do Serviço de Reconhecimento e Ordenamento Agrário', scale 1:25 000. SROA (1958).

'Carta de Uso e Ocupação do Solo de Portugal Continental (COS'90 e COS'07 - Nível 2)', scale 1:25 000. DGT (1990, 2007).

Fig. (6). Predicted LULC map for year 2024 for sector 1, 2 and 3.

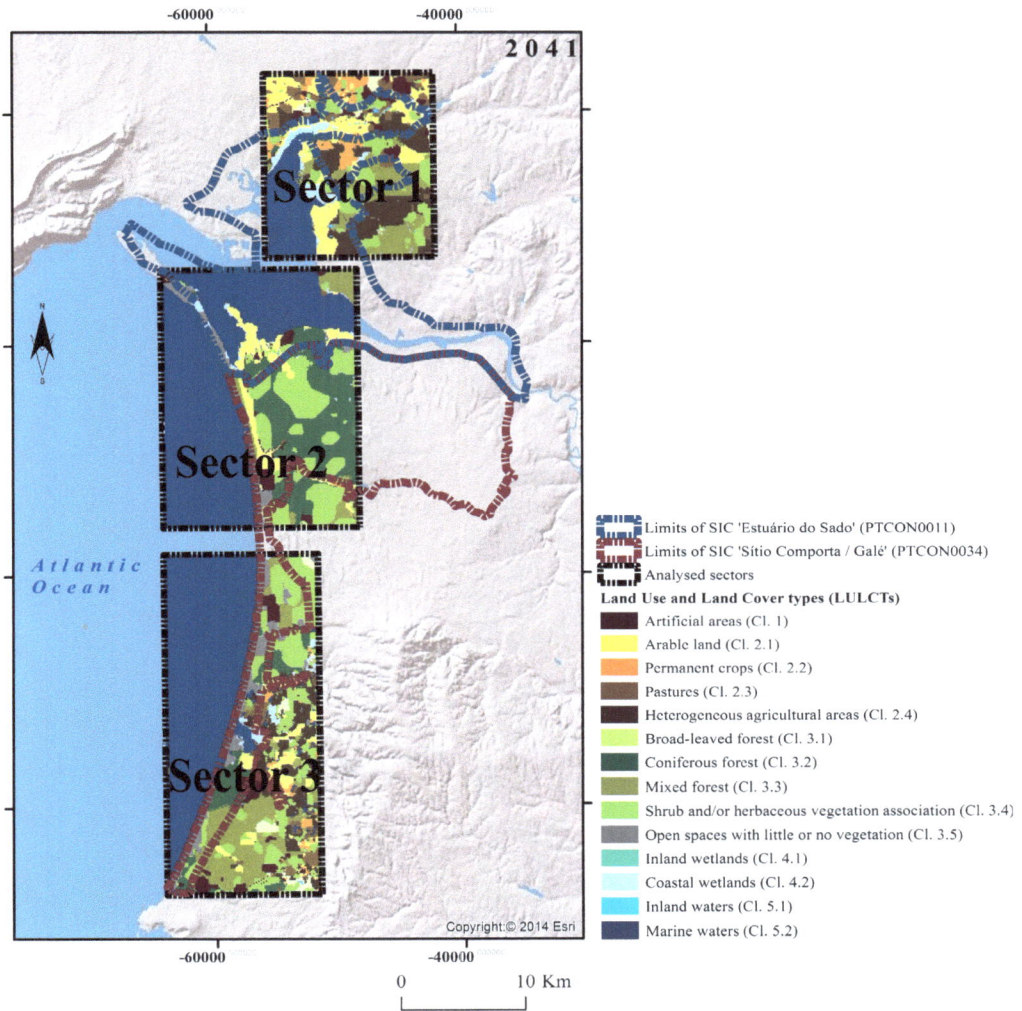

European Terrestrial Reference System 1989 (PT-TM06/ETRS89)

Source:
'Carta Agrícola e Florestal de Portugal do Serviço de Reconhecimento e Ordenamento Agrário', scale 1:25 000. SROA (1958).

'Carta de Uso e Ocupação do Solo de Portugal Continental (COS'90 e COS'07 - Nível 2)', scale 1:25 000. DGT (1990, 2007).

Fig. (7). Predicted LULC map for year 2041 for sector 1, 2 and 3.

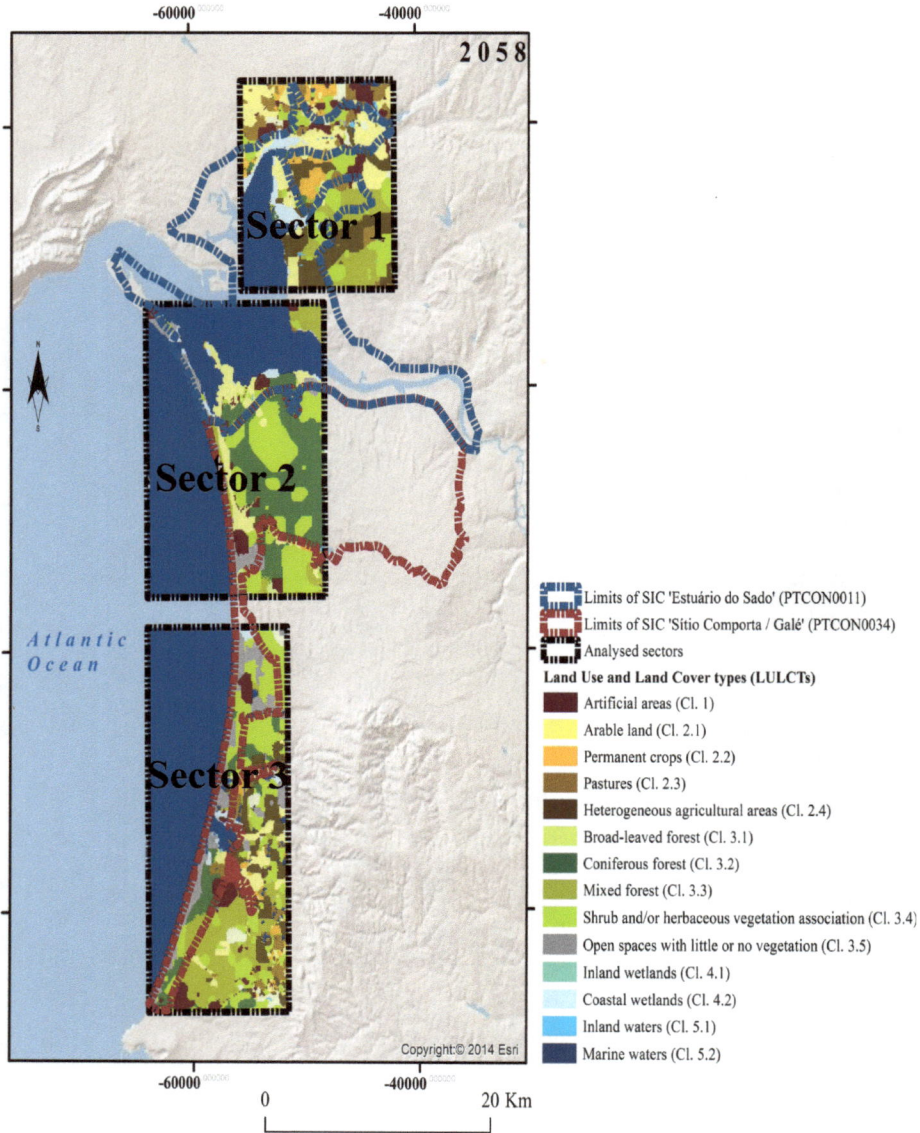

European Terrestrial Reference System 1989 (PT-TM06/ETRS89)

Source:
'Carta Agrícola e Florestal de Portugal do Serviço de Reconhecimento e Ordenamento Agrário', scale 1:25 000. SROA (1958).

'Carta de Uso e Ocupação do Solo de Portugal Continental (COS'90 e COS'07 - Nível 2)', scale 1:25 000. DGT (1990, 2007).

Fig. (8). Predicted LULC map for year 2058 for sector 1, 2 and 3.

Landscape Dynamics of Sector 1

In sector 1, the results obtained to Artificial areas and Agricultural and Agro-Forestry classes point out that the area occupied by the Artificial areas (Cl. 1), Arable land (Cl. 2.1), Permanent crops (Cl. 2.2) and Pastures (Cl. 2.3) will increase by the year 2058. In the case of Heterogeneous agricultural areas (Cl. 2.4) that occupy only 15% of the territory, whereas in 2007 they accounted for 25% (Fig. **9**).

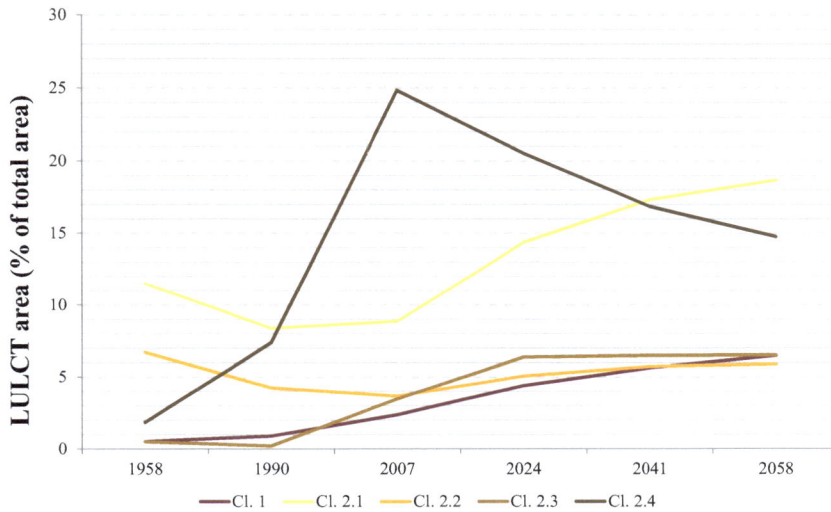

Fig. (9). Past and future trends of Artificial areas and Agricultural and Agro-Forestry classes in sector 1, 1958-2058 (% of total area).

Forest and natural/semi-natural areas shows a trend towards decline in Broad-leaved forest (Cl. 3.1), Coniferous forest (Cl. 3.2) and Mixed forest (Cl. 3.3); an increase of the Shrub and/or herbaceous vegetation association (Cl. 3.4) and the open spaces with little or no vegetation will be maintained (Cl. 3.5) (Fig. **10**).

Regarding wetland and marine bodies, in Coastal wetlands (Cl. 4.2) can be observed a clear regressive trend until the year 2058, while the Marine waters (Cl. 5.2) show a significant increase (Fig. **11**).

Analysis of the three previous figures reveals that if the changes that occurred between 1990 and 2007 persist into the future, the system would stabilize around the year 2058, with a predominance of wetland and marine bodies (20%), Arable land (19%), Heterogeneous agricultural areas (15%), forest (14%) and Shrub and/or herbaceous vegetation association (14%).

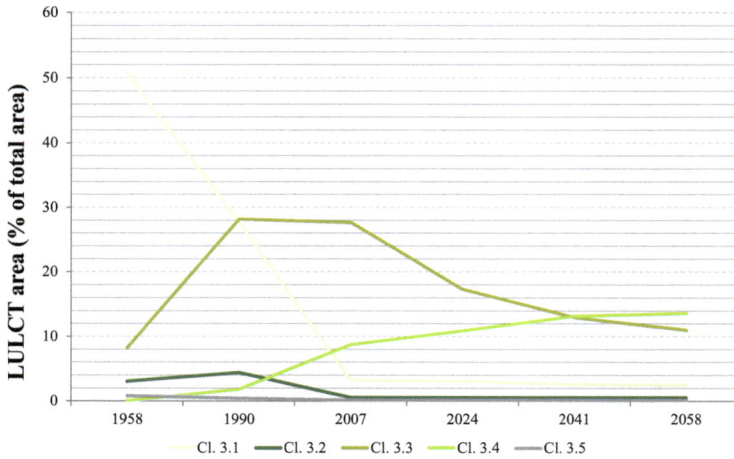

Fig. (10). Past and future trends of Forest and natural/ semi-natural classes in sector 1, 1958-2058 (% of total area).

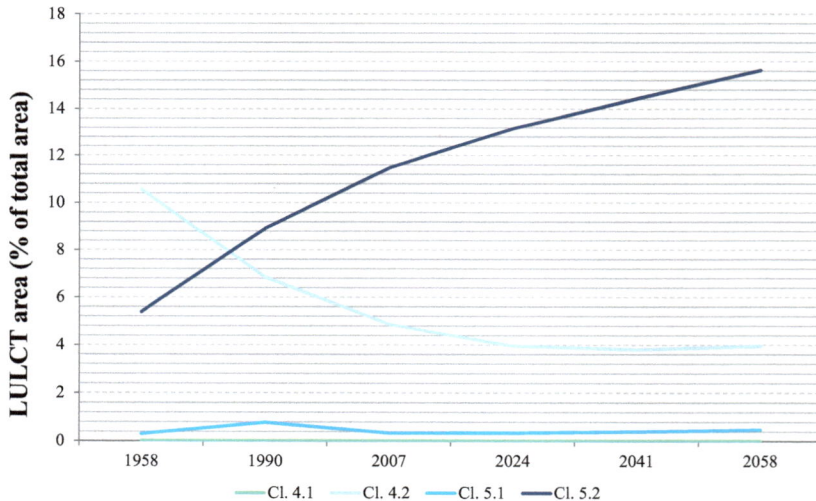

Fig. (11). Past and future trends of wetland and water bodies in sector 1, 1958-2058 (% of total area).

LULC Transition Analysis of Sector 1

Analyses of Table **2** shows that this sector, between 2007 and 2058, presents a total value of LULC changes of 43.2%. The increase of Cl. 2.1 (114.5%), Cl. 2.2 (57.9%) and Cl. 2.3 (95.2%) is due to decrease of Cl. 2.4 (-42.7%). The Cl. 1 will increase 169% mostly due to the decrease of Cl. 2.4, Cl. 3.3 e Cl. 3.4. However, the Cl. 3.4 will grow 54.5% due the conversion of Cl. 2.4, Cl. 3.1 e Cl. 3.3. It is also interesting to note that the decrease in Cl. 4.2 (-14.5%) will be in favor of Cl. 2.1, Cl. 2.4, Cl. 3.3 and Cl. 3.4. Moreover, the increase of Cl. 5.2 (37%) will be due to the reduction of Cl. 4.2, Cl. 2.1, Cl. 2.3, Cl. 2.4 and Cl 3.4.

Table 2. The long-term Cross-classification table (2007-2058) of sector 1.

To 2058 / From 2007	Cl. 1	Cl. 2.1	Cl. 2.2	Cl. 2.3	Cl. 2.4	Cl. 3.1	Cl. 3.2	Cl. 3.3	Cl. 3.4	Cl. 3.5	Cl. 4.1	Cl. 4.2	Cl. 5.1	Cl. 5.2	TOTAL
Cl. 1	459	18	0	11	10	0	0	6	9	0	0	0	0	0	513
Cl. 2.1	22	1676	6	25	0	6	0	0	15	0	0	104	0	61	1915
Cl. 2.2	40	94	596	45	0	0	0	0	18	0	0	0	0	0	793
Cl. 2.3	20	38	0	495	35	0	0	3	63	0	0	39	0	57	750
Cl. 2.4	298	1333	356	367	2027	4	19	3	709	0	0	174	48	33	5371
Cl. 3.1	38	73	45	40	22	384	0	0	78	0	0	0	0	0	680
Cl. 3.2	0	0	3	2	6	0	105	0	3	0	0	0	0	0	119
Cl. 3.3	319	807	246	430	939	111	46	2302	675	0	0	95	0	0	5970
Cl. 3.4	184	69	0	49	37	22	0	35	1320	0	0	98	3	53	1870
Cl. 3.5	0	0	0	0	0	0	0	0	0	37	0	0	0	0	37
Cl. 4.1	0	0	0	0	0	0	0	0	0	0	0	0	0	0	**0**
Cl. 4.2	0	0	0	0	0	0	0	0	0	0	0	384	0	668	1052
Cl. 5.1	0	0	0	0	0	0	0	0	0	0	0	4	23	47	74
Cl. 5.2	0	0	0	0	0	0	0	0	0	0	0	1	0	2479	2480
TOTAL	1380	4108	1252	1464	3076	527	170	2349	2890	37	**0**	899	74	3398	21624
Δ %	*169,0*	*114,5*	*57,9*	*95,2*	*-42,7*	*-22,5*	*42,9*	*-60,7*	*54,5*	*0,0*	*0,0*	*-14,5*	*0,0*	*37,0*	*43,2*

Landscape Dynamics of Sector 2

Similarly to the sector 1, the artificial areas show an increasing trend until 2058. The results for this sector also point to a reduction in Agricultural and Agro-Forestry classes (Fig. **12**).

Concerning the Forest and natural/semi-natural areas it is important to mention the significant growth of the Shrub and/or herbaceous vegetation association (Cl. 3.4) with an estimated projection of increase from 11% in 2007 to 15% in 2058, and a variation of the same magnitude in the Coniferous forest (Cl. 3.2) (decreasing from 24% in 2007 to 17% in 2058) (Fig. **13**).

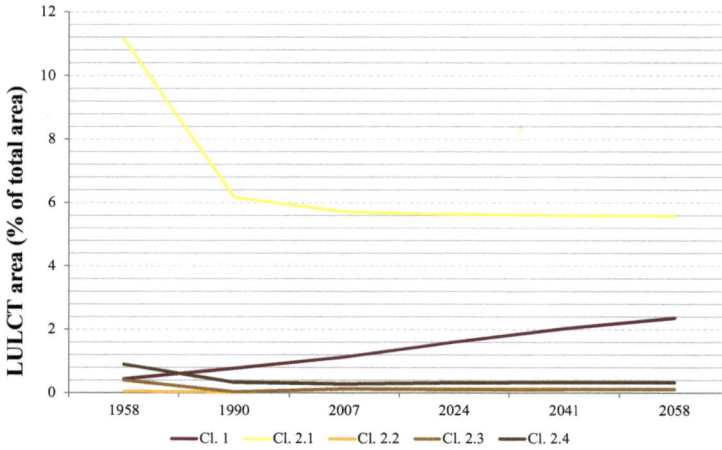

Fig. (12). Past and future trends of Artificial areas and Agricultural and Agro-Forestry classes in sector 2, 1958-2058 (% of total area).

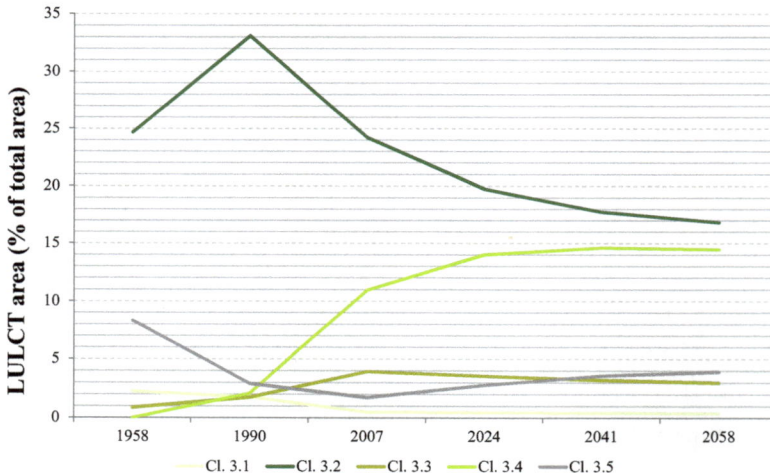

Fig. (13). Past and future trends of Forest and natural/ semi-natural classes in sector 2, 1958-2058 (% of total area).

Since 1958 the trends shown in wetland and marine bodies are similar to those in the previous sector (Fig. **14**).

This sector progresses to a stabilized forest landscape, around the year 2058, with 20% of forest, 15% of shrub lands, 6% of agricultural land, 4% of open spaces with little or no vegetation and 2% of artificial areas. Again, the wetland and marine bodies' classes will have the largest expression within the sector (53% of its area).

Fig. (14). Past and future trends of wetland and water bodies in sector 2, 1958-2058 (% of total area).

LULC Transition Analysis of Sector 2

The contiguity analysis between 2007 and 2058 for sector 2 presents a total value of LULC changes of 14.9% (Table **3**). One note is the increase of Cl. 2.4 (103%) due to the decrease of Cl. 2.1 and Cl. 3.2. Cl. 1 also increased 103%, mainly due to the reduction of Cl. 2.1, Cl. 3.2 and Cl. 3.4. However, Cl. 3.4 will increase 33.1% due the conversion of Cl. 3.2. Cl. 3.5 present a value of 126% due to the decrease of Cl. 2.1, Cl. 3.2, Cl. 3.3 and Cl. 3.4.

Table 3. The long-term cross-classification table (2007-2058) of sector 2.

To 2058 / From 2007	Cl. 1	Cl. 2.1	Cl. 2.2	Cl. 2.3	Cl. 2.4	Cl. 3.1	Cl. 3.2	Cl. 3.3	Cl. 3.4	Cl. 3.5	Cl. 4.1	Cl. 4.2	Cl. 5.1	Cl. 5.2	TOTAL
Cl. 1	395	0	0	0	0	0	0	0	1	0	0	0	0	0	**396**
Cl. 2.1	108	1660	0	0	51	0	0	46	0	46	0	51	0	51	**2013**
Cl. 2.2	0	0	45	0	0	0	0	0	0	0	0	0	0	0	**45**
Cl. 2.3	0	0	0	43	0	0	0	0	0	0	0	0	0	0	**43**
Cl. 2.4	0	0	0	0	99	0	0	0	0	0	0	0	0	0	**99**

(Table 3) contd.....

To 2058 / From 2007	Cl. 1	Cl. 2.1	Cl. 2.2	Cl. 2.3	Cl. 2.4	Cl. 3.1	Cl. 3.2	Cl. 3.3	Cl. 3.4	Cl. 3.5	Cl. 4.1	Cl. 4.2	Cl. 5.1	Cl. 5.2	TOTAL
Cl. 3.1	0	0	0	0	0	166	0	0	5	0	0	0	0	0	171
Cl. 3.2	146	71	0	0	51	43	5648	90	1922	410	0	100	0	49	8530
Cl. 3.3	46	0	0	0	0	0	1	1015	265	53	0	0	0	0	1380
Cl. 3.4	109	27	0	0	0	13	178	23	2935	330	0	137	0	103	3855
Cl. 3.5	0	0	0	0	0	0	0	2	2	533	0	11	0	59	607
Cl. 4.1	0	0	0	0	0	0	0	0	0	0	7	0	0	0	7
Cl. 4.2	0	0	0	0	0	0	0	0	0	0	0	19	0	656	675
Cl. 5.1	0	0	0	0	0	0	0	0	0	0	0	0	42	0	42
Cl. 5.2	0	0	0	0	0	0	0	0	0	0	0	0	0	17371	17371
TOTAL	804	1758	45	43	201	222	5827	1176	5130	1372	7	318	42	18289	35234
Δ %	103,0	-12,7	0,0	0,0	103,0	29,8	-31,7	-14,8	33,1	126,0	0,0	-52,9	0,0	5,3	14,9

As in the previous sector, there is a tendency to decrease in Cl. 4.2 (-52.9%) in favor of Cl. 2.1, Cl. 3.2 and Cl. 3.4. The slight increase observed in Cl. 5.2 (5.3%) is mainly due to the decrease of Cl. 4.2 and Cl 3.4.

Landscape Dynamics of Sector 3

This sector continues to present a trend for an urban sprawl by 2058. Agricultural and agro-forestry classes, with the exception of Arable land (Cl. 2.1) is forecast to have a particular expression in the territory of this sector (Fig. **15**).

On the other hand, Broad-leaved forest (Cl. 3.1), Coniferous forest (Cl. 3.2) and Mixed forest (Cl. 3.3) tend to suffer a gradual reduction between 2007 and 2058. In the same period, the Shrub and/or herbaceous vegetation association (Cl. 3.4) reveals a slight increase and subsequent stabilization of its occupation area. In this time period, Open spaces with little or no vegetation (Cl. 3.5) present the most interesting pattern of this sector, where in 2007 the class represents 2.4% of the total sector and is expected that in 2058 will take approximately 8% (Fig. **16**).

Wetland and marine bodies will increase again until 2058 with an estimated projection of 42% for Marine waters (Cl. 5.2) and 1% for Coastal wetlands (Cl. 4.2) (Fig. **17**).

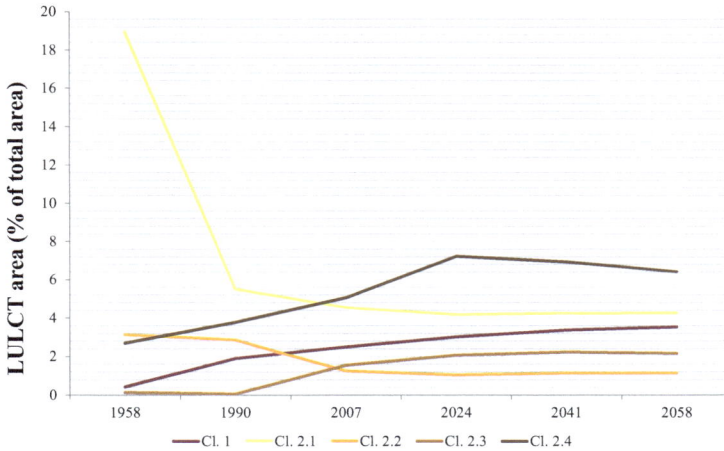

Fig. (15). Past and future trends of Artificial areas and Agricultural and Agro-Forestry classes in sector 3, 1958-2058 (% of total area).

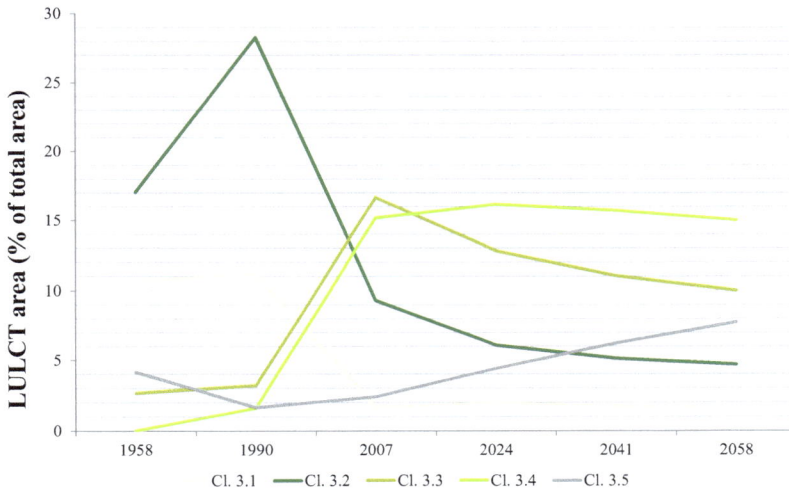

Fig. (16). Past and future trends of Forest and natural/ semi-natural classes in sector 3, 1958-2058 (% of total area).

LULC Transition Analysis of Sector 3

In sector 3 it was observed a total value of LULC changes of 24.3% (Table **4**). As in previous sectors continues to verify the increase of Cl. 1 (39.2%). It is interesting to analyze the maintenance of Cl. 3.4 and the decrease of Cl. 3.2 (-51.2%) and Cl. 3.3 (-41.2%). However, as in sector 2, Cl. 3.5 has a value of 227.1% due to the significant decrease of the Cl. 3.4.

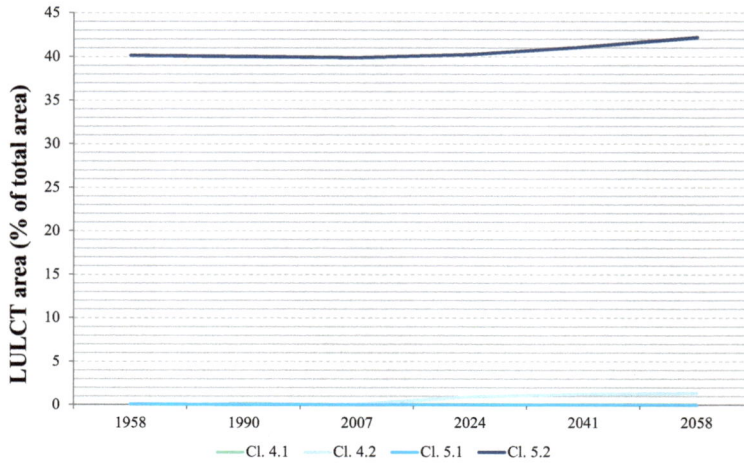

Fig. (17). Past and future trends of wetland and water bodies in sector 3, 1958-2058 (% of total area).

Table 4. The long-term cross-classification table (2007-2058) of sector 3.

To 2058 / From 2007	Cl. 1	Cl. 2.1	Cl. 2.2	Cl. 2.3	Cl. 2.4	Cl. 3.1	Cl. 3.2	Cl. 3.3	Cl. 3.4	Cl. 3.5	Cl. 4.1	Cl. 4.2	Cl. 5.1	Cl. 5.2	TOTAL
Cl. 1	810	1	0	46	2	0	2	10	9	30	0	0	0	0	910
Cl. 2.1	1	1218	10	64	61	4	0	1	42	46	0	47	0	151	1645
Cl. 2.2	0	64	238	48	52	0	1	0	36	6	0	2	0	2	449
Cl. 2.3	0	0	0	152	0	0	0	0	0	0	0	0	0	410	562
Cl. 2.4	4	103	47	113	1257	21	0	2	55	78	0	71	0	91	1842
Cl. 3.1	46	3	0	5	18	407	0	0	71	39	0	2	0	0	591
Cl. 3.2	104	29	6	98	279	51	1521	155	762	322	0	48	0	15	3390
Cl. 3.3	191	176	96	277	517	85	0	3021	1036	425	0	201	0	39	6064
Cl. 3.4	111	0	1	116	115	74	132	378	3510	1036	0	37	0	12	5522
Cl. 3.5	0	0	0	0	0	0	0	0	1	875	0	0	0	0	876
Cl. 4.1	0	0	0	0	0	0	0	0	0	0	25	0	0	0	25
Cl. 4.2	0	0	0	0	0	0	0	0	0	0	0	30	0	0	30
Cl. 5.1	0	0	0	0	0	0	0	0	0	0	0	0	7	0	7
Cl. 5.2	0	0	0	0	0	0	0	0	0	8	0	0	0	14535	14543
TOTAL	1267	1594	398	919	2301	642	1656	3567	5522	2865	25	438	7	15255	36456
Δ %	39,2	-3,1	-11,4	63,5	24,9	8,6	-51,2	-41,2	0,0	227,1	0,0	1360,0	0,0	4,9	24,3

Similarly, it is quite remarkable the considerable increase of Cl. 4.2 in this sector, as consequence of the expected expansion of these wetlands to inland territories that were mostly occupied by Cl. 2.1, Cl. 2.4, Cl. 3.2, Cl. 3.3 and Cl. 3.4. This is also justified by the increase of Cl. 5.2 in 2058 (4.9%).

The LULC changes related to Cl. 3.5, Cl. 4.2 and Cl. 5.2, forecast to occur between the period 2007 and 2058, may indicate the occurrence of a new transgressive phase similar to that described by Mateus [37].

DISCUSSION

LULC changes in the three sectors within the 'Sado Estuary' and 'Comporta-Galé' Sites of Community Importance will continue to represent the interaction between people and territory, requiring more integrated studies on landscape dynamics.

Several factors (biotic, abiotic and public and private policies) acting separately, but multiplying their effects when acting together, make the subsequent evaluation of results extremely complex and difficult. Usually they are generated by non-linear interactions among the system components. Although the results do not accurately define the factors that are seen as being responsible for these changes, this study allows for a better understanding of the past, present and future dynamics of the mosaic of ecosystems.

In general can be assume that these sectors, which are not far from Lisbon, evolved so that heath lands and shrub lands (described by Mateus [37]) were substituted by agriculture and forest in the nineteenth century, and in the 1960s and 1970 were occupied by urban areas (social areas) as result of the expansion of the Metropolitan Area of Lisbon. In the course of time, agriculture area decreased and forest-land cover increased.

Throughout the first half of the twentieth century was lead significant economic and political changes that influenced Portuguese agricultural policies. The purpose of these policies was to intensify the exploitation of uncultivated lands by agriculture and forestry. The main concern about the short-age of cereals (particularly wheat) conducted to the "Wheat Campaign" in 1929 and the culture of other cereals (rice, rye and corn). This originated a great transformations in the landscape, mainly in the 1930s and 1940s when the "Forestry Plan" was implemented. To the south of the River Tagus, the afforestation plan had, as its resolutions, the development and improvement of cork oak plantations for the cork industry. At the end of the 1950s and in the 1960s (corresponding to the "dawn of the contemporary industrialization") larger plantations of Eucalyptus and Maritime Pine stands covered large landscape zones for the cellulose industry

[38]. It was also observed that the urban expansion of the Metropolitan Area of Lisbon, caused by the bridge construction in 1966, which connected Lisbon to the Setúbal Peninsula, led to significant changes in the land occupation (because the growing population demands and urban expansion). With the 1974 Revolution and its consequences in the agricultural reforms, the incorporation of Portugal in the European Community in 1986 was lead by the Common Agriculture Policy (CAP). Thus in these decades, at the LULC level, the major significances were the reduction of agricultural areas in favor of urban and forest areas. However, there has been a great demand for sandy soils by intensive agriculture (in the last twenty years) which puts great pressure on the coastal natural areas which exhibit the most important Natura habitats and flora in Mainland Portugal [39 - 41].

During the 1990s in order to protect sensitive areas and others with potential for agriculture also emerged various spatial planning policies, such as the municipal master plan (PDM), special plans, National Ecological Reserve (REN), National Agricultural Reserve (RAN), National Network of Protected Areas and sectoral plans [42].

It should be noted that, especially in sectors 2 and 3, the tourism activity gained significant importance and has been playing to the present an important role as an economic activity. However, the appearance in 2007 of the National Planning Program (PNPOT) aimed to avoid a scattered occupation, and promote a growing integration of environmental concerns and sustainability and also the revision of the PDM.

Aside from sector 1 (between 1990 and 2007), the agricultural areas decreased substantially, which is directly related to the markets and policies, in particular the mentioned above CAP.

The exception regime of the sector 1 was related to the CAP in 1992, with the inclusion of a package of agri-environmental measures, namely the encouragement of farmers to abandon the agricultural and forest areas and to not cultivate its lands for at least 20 years in order to promote more compatible uses with the environment and the maintenance of rural characteristics [43].

Regarding the forest areas, mainly in the sector 2 and 3, increased in the period 1958-1990, due to markets, EU funds and the profitability of the Portuguese forest [43].

In the Pinus forests was observed a decrease between 1990 and 2007; however the two main species included in this class revealed a distinct behaviour. The maritime pine forests decreased due to the low added value of their stocks, they are replaced by eucalyptus (lower demand management and increased market demand), and the dissemination of the pine wood nematode and consequent cutting of pine trees. On the other hand, the stone pine areas increased due to measures to support the planting of agricultural land promoted by public policies and premiums paid under their planting.

Concerning the wetlands (includes salt marshs habitats) and marine environments it should be noted the in the Sado Estuary case, the last decades revealed a reduction of the salt marsh habitats, which led to the erosion and consequent extinction of large patches of low marshes [22]. Three factors are responsible for the salt marsh retreat, the long term and possible future erosion of the Sado estuary salt marshes: directly by anthropogenic activities (fishing activities, shellfish harvesting, earthworm collecting for sea fishing, grazing, among others); sedimentation and indirectly by the sea level rise [44].

In this sense, we observed that the LULC changes between 1958-2007 is a Markovian process, and it is possible to forecast LULC, assuming the above mentioned historical changes are stationary over time. For sector 1 the CA LULC changes Model predict in 2058 the increase of Artificial areas, Agricultural classes (except the Heterogeneous agricultural areas), Coniferous forest, Shrub lands and Marine waters. By contrast, the model forecast a decrease of Broad-leaved and Mixed forests and Wetlands. In the latter case, the projection reflects a gradual erosion trend. In fact, the transition probabilities show an evident rate of decrease until 2024. Only in 2058 will be expected to remain relatively stable. The great reduction of salt marshes is clear and will continue to be affected by the sea level rise.

For sector 2 the model highlight the increase of Artificial areas, Heterogeneous agricultural areas, Broad-leaved forest, Shrub lands, Open spaces with little or no vegetation and Marine waters. On the other hand, the model predicts the decrease of Arable land, Coniferous and Mixed forests and Wetlands.

In the case of sector 3, the model forecast an increase of Artificial areas, Pastures, Heterogeneous agricultural areas, Broad-leaved forest, Open spaces with little or no vegetation, Wetlands and Marine waters. On the other hand, the model point out a decrease of Arable land, Permanent crops, Coniferous and Mixed forests.

However, the stochastic modeling (MCs) deserves some reservations about its use, because the cyclic economic factors, natural disasters (such as forest fires), or specific development policies can compromise their conditional character. Also not taken account the spatial dimension, and therefore, makes the CA modeling crucial for the spatial distribution of the forecast scenarios.

The GIS based modeling approach proved to be an important spatial planning tool, allowing to understand the present interactions and their connections with the past, and contains the great advantage of producing trends for the future, which cannot be completely exact, they are a reflection of the landscape dynamics.

CONCLUSION

This chapter shows how GIS and stochastic modeling techniques can explain complex spatial processes. The proposed approach integrating MCs, CA and GIS proved to be a valuable tool for planners so as to better understand and anticipate the impact and magnitude of LULC changes. MC analysis and CA simulation can be used as a heuristic tool to response questions such as "what happens if...?", considering the constancy of the transition probabilities over time. Regarding the quantification of the fragmentation processes and LULC changes in 'Sado Estuary' and 'Comporta-Galé' Natura 2000 Sites, these models are able to reveal non-obvious trends in the data and to describe ecological patterns.

From an applied research point of view, this approach is useful to identifying adequate planning and management approaches for coastal ecosystems, for monitoring and planning natural and protected environments.

CONSENT FOR PUBLICATION

Not applicable.

CONFLICT OF INTEREST

The author (editor) declares no conflict of interest, financial or otherwise.

ACKNOWLEDGEMENTS

The authors would like to thank Foundation for Science and Technology (FCT) (PhD Project «Structure and dynamics of habitats and landscapes of Sado Estuary and Comporta/Galé Places»/ SFRH/BD/45147/2008) for their funding of this research.

REFERENCES

[1]   P. Arsénio, *"Qualidade da Paisagem e Fitodiversidade. Contributo Para o Ordenamento e Gestão de Áreas Costeiras de Elevado Valor Natural"*, PhD dissertation. Instituto Superior de Agronomia: Universidade Técnica de Lisboa, 2011.

[2]   F. Gutierres, *"Structure and dynamics of habitats and landscape of Sado Estuary and Comporta/Galé Natura 2000 Sites - A contribution to sustainable land management and ecological restoration"*, PhD dissertation. Institute of Geography and Territorial Planning: Universidade de Lisboa, 2014.

[3]   J. Candau, and N. Goldstein, "Multiple scenario urban forecasting for the California south coast region", *Proceedings of the 40th Annual Conference of the Urban and Regional Information Systems Association,* Chicago, Illinois.

[4]   J. Cheng, *Modeling Spatial & Temporal Urban Growth. Faculty of Geographical Sciences.* Utrecht University: Utrecht, 2003, p. 203.

[5]   P. Cabral, J.P. Gilg, and M. Painho, "Monitoring urban growth using remote sensing, GIS and spatial metrics", *Proceedings of SPIE Optics & Photonics: Remote sensing and modeling of ecosystems for sustainability,* 2005, pp. 1-9, San Diego, USA.
        [http://dx.doi.org/10.1117/12.614852]

[6]   I. Benenson, and P. Torrens, *Geosimulation: Automata-based modeling of urban phenomena.* Wiley, 2004, pp. 1-312.
        [http://dx.doi.org/10.1002/0470020997]

[7]   K. Clarke, "A decade of Cellular Urban Modelling with SLEUTH: Unresolved Issues and Problems", *Planning Support Systems for Cities and Regions,* vol. 3, pp. 47-60, 2008.

[8]   Y. Murayama, and R.B. Thapa, *Spatial Analysis and Modeling in Geographical Transformation Process (GIS-based Applications).* Springer: Netherlands, 2011, pp. 1-302.
        [http://dx.doi.org/10.1007/978-94-007-0671-2]

[9]   Q. Weng, "Land use change analysis in the Zhujiang Delta of China using satellite remote sensing, GIS and stochastic modelling", *J. Environ. Manage.,* vol. 64, no. 3, pp. 273-284, 2002.
        [http://dx.doi.org/10.1006/jema.2001.0509] [PMID: 12040960]

[10]   P. Godinho-Ferreira, M. Almeida, and A. Fernandes, Landscape Dynamics in the Area of Serra da Arrábida and the Sado River Estuary. *In: Recent Dynamics of the Mediterranean Vegetation and Landscape.,* S. Mazzoleni, G. di Pasquale, M. Mulligan, Eds., John Wiley and Sons: Chichester, 2004, pp. 201-209.
        [http://dx.doi.org/10.1002/0470093714.ch17]

[11]   W.H. Romme, and D.H. Knight, "Landscape diversity: the concept applied to Yellowstone National Park", *Bioscience,* vol. 32, pp. 664-670, 1982.
        [http://dx.doi.org/10.2307/1308816]

[12]   D.L. Urban, "Landscape ecology: a hierarchical perspective can help scientists understand spatial patterns", *Bioscience,* vol. 37, pp. 119-127, 1987.
        [http://dx.doi.org/10.2307/1310366]

[13]   P.M. Torrens, *How Cellular Models of Urban Systems Work. WP-28, Report, Centre for Advanced Spatial Analysis.* University College London: London, 2000.

[14]   P.M. Torrens, "Simulating Sprawl", *Ann. Assoc. Am. Geogr.,* vol. 96, no. 2, pp. 248-275, 2006.
        [http://dx.doi.org/10.1111/j.1467-8306.2006.00477.x]

[15]   X. Yang, and C.P. Lo, "Modeling urban growth and landscape changes in the atlanta metropolitan area", *Int. J. Geogr. Inf. Sci.,* vol. 17, pp. 463-488, 2003.
        [http://dx.doi.org/10.1080/1365881031000086965]

[16]   J.F. De Raaf, H.G. Reading, and R.G. Walker, "Cyclic sedimentation in the Lower Westphalian of North Devon, England", *Sedimentology,* vol. 4, no. 1-2, pp. 1-52, 1965.
        [http://dx.doi.org/10.1111/j.1365-3091.1965.tb01282.x]

[17]   C.H. Harper, Improved Method of Facies Sequences Analysis. *Facies Models, Geoscience Canada, Reprint Series* Miami Shores: FL, USA, 1984.

[18]   A.D. Miall, "Markov chain analysis applied to an ancient alluvial plain succession", *Sedimentology,* vol. 20, no. 3, pp. 347-364, 1973.
        [http://dx.doi.org/10.1111/j.1365-3091.1973.tb01615.x]

[19]   D.W. Powers, and R.G. Easterling, "Improved methodology for using embedded Markov chains to

describe cyclical sediments", *J. Sediment. Petrol.,* vol. 52, no. 3, pp. 913-923, 1982.

[20]   J. Rocha, *"Complex Systems, Modeling and Geosimulation of the Evolution of Land Use and Cover Patterns"*, PhD dissertation. Institute of Geography and Territorial Planning: Universidade de Lisboa, 2012.

[21]   S. Rivas-Martínez, *Memória del mapa de séries de vegetación de España 1: 400.000.* ICONA: Madrid, 1987.

[22]   J.C. Costa, P. Arsénio, and T. Monteiro-Henriques, "Finding the Boundary between Eurosiberian and Mediterranean Salt Marshes", *10th International Coastal Symposium,* 2009, pp. 1340-1344, Lisboa.

[23]   J.C. Costa, C. Neto, and C. Aguiar, "Vascular Plant Communities in Portugal (Continental, The Azores and Madeira)", *Global Geobotany,* vol. 2, pp. 1-180, 2012.

[24]   C. Neto, "A flora e a vegetaçao do superdistrito Sadense (Portugal)", *Guineana,* vol. 8, pp. 1-269, 2002.

[25]   C. Neto, M.E. Moreira, and R.M. Caraça, "Landscape Ecology of the Sado River Estuary (Portugal)", *Quercetea,* vol. 7, pp. 43-64, 2005.

[26]   A. Cancela d'Abreu, T. Pinto Correia, and R. Oliveira, *"Contributos para a Identificação e Caracterização da Paisagem em Portugal Continental",* Universidade de Évora: DGOTDU, 2004.

[27]   ArcGIS Desktop [Computer Program]. Version 10, *Redlands, CA: Environmental Systems Research Institute (ESRI),* 2011.

[28]   N. Guiomar, T. Batista, and J.P. Fernandes, *Corine Land Cover Nível 5 - Contribuição para a Carta de Uso do Solo em Portugal Continental.* Associação de Municípios do Distrito de Évora: Évora, 2009.

[29]   N.P. Hegde, I.V. MuraliKrishna, and K.V. ChalapatiRao, Integration of Cellular Automata and GIS for Simulating Land Use Changes. WG II - 7 -Quality of Spatio- Temporal Data and Models. *5th International Symposium Spatial Data Quality,* A. Stein, Ed., ITC: Enschede, 2007.

[30]   M. Kijima, *Markov Processes for Stochastic Modeling (Stochastic Modeling Series).* 1st ed. Chapman & Hall: London, 1997, pp. 1-341.
        [http://dx.doi.org/10.1007/978-1-4899-3132-0]

[31]   J.R. Eastman, *IDRISI Kilimanjaro: guide to GIS and image processing.* Clark University, Clark Labs: Worcester, 2003.

[32]   K.C. Clarke, S. Hoppen, and L. Gaydos, "A self-modifying cellular automaton model of historic urbanization in the san francisco bay", *Environ. Plann. B Plann. Des.,* vol. 24, pp. 247-261, 1997.
        [http://dx.doi.org/10.1068/b240247]

[33]   Y. Liu, *Modelling Urban Development with Geographical Information Systems and Cellular Automata. EUA.* CRC Press, 2008, pp. 1-186.
        [http://dx.doi.org/10.1201/9781420059908]

[34]   P. Cabral, S. Santos, A. Zamyatin, and M. Painho, Multi-decadal Projection and Analysis of Land Use and Cover Changes in mainland Portugal. *Direção-Geral do Território (DGT), Eds. LANDYN - Land use and cover change in Continental Portugal: characterization, driving forces and future scenarios,* 2014.

[35]   B.J. Murteira, and M. Antunes, *Probabilidades e Estatística.* vol. II. 2nd ed. Escolar Editora: Lisbon, 2012, pp. 1-612.

[36]   IDRISI [Computer Program]. Version TAIGA, Worcester, MA, USA: Clark Labs, Clark University.

[37]   J.E. Mateus, *"Holocene and present-day ecosystems of the Carvalhal region, South-west Portugal",* PhD dissertation. Utrecht, The Netherlands: University of Utrecht, 1992.

[38]   E. Castro Caldas, A agricultura portuguesa através dos tempos. *Série Sociologica-2* 1st ed. INIC: Lisboa, 1991.

[39]   J.C. Costa, C. Neto, and M. Martins, "Annual dune plant communities in the southwest coast of europe", *Plant Biosyst.,* vol. 141, no. 1, pp. 91-104, 2011.
[http://dx.doi.org/10.1080/11263504.2011.602729]

[40]   M. Martins, C. Neto, and J.C. Costa, "The meaning of mainland Portugal beaches and dunes' psammophilic plant communities: a contribution to tourism management and nature conservation", *J. Coast. Conserv.,* vol. 17, no. 3, pp. 279-299, 2013.
[http://dx.doi.org/10.1007/s11852-013-0232-9]

[41]   C. Neto, J.C. Costa, and J. Honrado, "Phytosociological associations and Natura 2000 habitats of Portuguese coastal dunes", *Fitosociologia,* vol. 44, no. 2, suppl. Suppl. 1, pp. 29-35, 2008.

[42]   M. Costa Lobo, *Planeamento Urbanístico em Portugal, in on the w@terfront.* Public Art and Urban Design da Universitat de Barcelona, 2011.

[43]   B. Condessa, I.L. Ramos, and M.G. Saraiva, Identificação das Principais Forças Motrizes em Termos de Políticas Públicas na Alteração da Ocupação do Solo em Portugal Continental. *Direção-Geral do Território (DGT), Eds. Uso e Ocupação do Solo em Portugal Continental: avaliação e cenários futuros. Projeto LANDYN. Lisboa: DGT,* 2014, pp. 63-80.

[44]   D. Almeida, C. Neto, L. Esteves, and J.C. Costa, "The impacts of land-use changes on the recovery of salt marshes in Portugal", *Ocean Coast. Manage.,* vol. 92, pp. 40-49, 2014.
[http://dx.doi.org/10.1016/j.ocecoaman.2014.02.008]

CHAPTER 2

Agent-based Modelling of Tourists' Destination Decision-making Process

Inês Boavida-Portugal[1,*], Jorge Rocha[1], Carlos Cardoso Ferreira[2] and José Luís Zêzere[2]

[1] *Department of Spatial Planning and Environment, Faculty of Spatial Sciences, University of Groningen, Groningen, The Netherlands*

[2] *Institute of Geography and Spatial Planning, Universidade de Lisboa, Lisboa, Portugal*

Abstract: Agent-Based Models (ABM) are becoming more relevant in computational social science (CSS) due to the potential to model complex phenomena that emerge from individual-based interactions. Most tourism theoretical models recognize the complex nature of the tourism system, and complexity is a subject of growing interest among researchers. Geosimulation models (GM) are presented as potential tools to address tourism in a complex systems lens. Particularly ABM, has a GM tool, as captured growing interest by tourism researchers, however there is little empirical application as a tool to explore and predict tourism patterns. The purpose of the chapter is to frame ABM in GM following a complex systems theoretical approach, in order to increase knowledge by (i) considering the complex nature of tourism, (ii) providing tools to explore the interactions between system components, (iii) discussing the potential for coupling ABM and Geographical Information Systems (GIS) in tourism research, and (iv) giving insights on the functioning of the tourist behaviours and decision-making process through an ABM approach. Also a theoretical ABM is developed to improve knowledge on tourist decision-making in the selection of a destination to vacation. Tourists' behaviour, such as individual motivation and social network influence in the vacation decision-making process are presented. On-going work on loose coupling of ABM and GIS is discussed.

Keywords: Agent-Based Models, Cellular Automata, Complexity, Computational social science, Decision-making process, Distribution patterns, Exploratory analysis, Geosimulation, Geographical Information Systems, Heterogeneity, If-then rules, Individual-based, Interaction, Non-linearity, Scenario development, Simulation, Tourism, Tourist behaviour, Tourism system.

* **Corresponding author Inês Boavida-Portugal:** Department of Spatial Planning and Environment, Faculty of Spatial Sciences, University of Groningen, Groningen, The Netherlands; Tel: +31 50 36 36910; E-mail: i.boavida.portugal@rug.nl

Ana Cláudia Teodoro (Ed.)

INTRODUCTION

Tourism is a geographical explicit phenomenon that encompasses the movement of people, for leisure related purposes, between origin and destination(s). As Pearce [1] states movement is the basic element of tourism, and no other discipline concentrates on spatial patterns of tourism phenomena as geography [2]. Tourism geography, or the geography of tourism, began to emerge more consistently in academic journals in the second half of the 20th century, with focus on what Pearce calls the major areas of interest: spatial patterns, geography of resorts, movements and flows, tourism impacts, and tourist space models. Lew [3] argues that geographers are interested in factors that create, shape and influence places and space: i) the tourist, behaviour and experience; ii) the social, economic and political processes; iii) and the resulting geographical impacts that these combined produce.

Tourism geography research has mainly focused on the how, why and specially the where. In this context, Geosimulation models (GM) [4] provide tools for exploring geographical phenomena, *i.e.* spatially-related automata, such as tourism. GM result from the need to associate autonomous agents taking an Object-Oriented Programming (OOP) approach with geographic data or their spatial locations. This was possible through the symbiosis between Geographic Information Systems (GIS) and Agent-Based Models (ABM):

a. GIS and remote sensing databases serve as the information source.
b. System theory and complexity theory provide the theoretical paradigm and analytical tools for investigating GM.
c. ABM allow to simulate individual agents' (inter)actions providing the ability to model the emergence of new phenomena or behaviours [5] and evaluate the resulting system behaviour over space and time [6]. ABM are a relatively recent approach for simulating complex systems composed of interacting, heterogeneous and autonomous entities, also referred to as agents [7].

There is growing cross-disciplinary interest in the integration of GIS and ABM [4, 8, 9] and some examples of recent applications are for example: the simulation of pedestrians' movements [10], crowd congestion modelling [11], households' dynamics [12], urban growth [13], gentrification, and traffic simulation [14]. Readers can refer to Castle and Crooks [15] for a detailed review. Crooks [16] identifies the advantages of the coupling of GIS and ABM, on one hand for agent-based modellers that gain the ability to have geographical referenced agents, and on the other hand for GIS users it provides the ability to model the emergence of phenomena through agents individual interactions on a GIS over time and space.

The current work reports a complex systems approach to the comprehensive investigation of the tourist decision-making process in the selection of a destination. Where and why tourists decide to vacation form the main objective of the presented ABM. The paper structure starts with the exploration of complex systems theory and its characteristics, the value of coupling ABM and GIS in tourism research, and ABM applications in tourism. The chapter follows the reasoning from O'Sullivan and Haklay [17] starting from a broader scale of whether the "world is agent-based?" growing to is "tourism agent-based?". Then a theoretical ABM is developed to enhance the understanding of the tourist decision-making process, considering Kennedys' [18] proposition of: (i) individual motivations; (ii) rationality based on if–then rules; (iii) human emotions and satisfaction and (iv) the influence of the social network in the destinations choice. Scenario experiments were developed based on theoretical assumptions and results are briefly discussed. Also future and on-going work regarding the adaptation of the tourism system supply to demand trends using a loose coupling ABM – GIS/Cellular Automata (CA) approach is discussed.

ABM AND COMPLEX SYSTEMS

As a GM tool ABM implementation started as a set of tools and techniques to develop computational models of complex adaptive systems in physical and natural sciences [19]. As a matter of fact these models have been widely applied in biological sciences to simulate for instance cell behaviour and interaction, the immune system, and epidemiology. Ecology is also a pioneering research field where applications of ABM were developed to experiment with diverse populations of individuals and their interactions, such as the flocking of birds model [20, 21] and the predator-prey model [22, 23]. More recently, ABM are becoming a current approach in social systems simulation, highly leveraged by the development of computational social science (CSS) in the second half of the 20th century. ABM as a CSS tool can be broadly categorized in five areas: automated information extraction, social network analysis, GIS, complex systems and social simulations models [24].

Besides the importance that CSS has had in ABM approach spread in social systems simulation, the theoretical change of perspective from a reductionist approach to a complex systems paradigm is also relevant. Following the reductionist approach, based on the Newtonian paradigm of scientific inquiry, systems are understood to be highly dependent on initial conditions that are paramount to future outcomes [25]. Systems assume clockwork machine processes that can be fully explained by its individual components and behaviours, thus operating in a linear and predictable way. However, in real-world systems simple cause and effect relationships rarely exist and instead very small changes can produce large effects on the system, referring to Lorenz's butterfly effect theory [26]. Systems characterized by disorder, instability and non-linearity are predominant and considered complex. Through the complex systems approach it is suggested that social systems, alike ecological systems [27], are highly unpredictable and non-linear, emerging from the bottom-up, individually based heterogeneous behaviours, and following an interaction-by-interaction approach.

However, the concepts and definitions central to this area of scientific inquiry remains elusive at both the qualitative and quantitative levels. There are still many mix-ups over the terminology "complexity" which is often misused and semantically mistaken with the concept "complicated" across research fields, resulting in what Scott (in Eidelson [28]) calls an "interdisciplinary Tower of Babel". Systems are complicated when they can be fully explained by its individual components, which are typically many. Chapman (in Reitsma [29]) states that "complication is a quantitative escalation of that which is theoretically reducible". This concept falls in line with the above mentioned reductionist approach.

In fact, many systems appear simple, but reveal remarkable complexity when closely examined. Others appear complex, but can be described simply, like for instance some machines. Complex systems cannot be fully understood by analysing its individual components but rather by the interactions between them and emerging behaviours that rise from the bottom-up [30]. A complex system can only be understood by considering it as a whole, almost independently of the number of parts composing it [31]. A "simple" flock of birds, bees and ants working societal structure, the behaviour of consumers in a retail environment, people and groups in a community, the economy, the weather, traffic jams, the immune system, are some well-known examples of complex systems able to adapt to the external conditions without apparently recognizable organization but following a number of simple individually based rules [32 - 35].

Perhaps one should not be surprised if complexity cannot be given a simple definition. There are several definitions of complex systems and some of the main characteristics are identified by authors as: non-determinism, non-linearity, emergence and self-organisation, presence of feedback loops, distributed nature, self-similarity, limited decomposability, local interaction, oscillations between equilibrium and chaotic state [30, 31, 36, 37]. These concepts are briefly discussed below:

i. Complex systems are non-deterministic in a way that is impossible to anticipate precisely its behaviour even based on differential equations of all its elements. Therefore predictions are usually probabilistic.

ii. Interactions are mathematically non-linear, meaning that disruptions or perturbations in one component may have disproportional effect in other components [7]. Taking on the example of Lorenz's butterfly effect theory, which defended that a wing flapping of a butterfly could initiate a series of effects resulting in a cyclone [26], one is confronted with the hypothesis of a very small change producing unexpected large effects on the system, reshaping it in new ways. The opposite can also happen, and often effects thought to have great impact may actually not result in any changes on the overall system [38]. It is a requirement for complexity [30].

iii. The described non-linear behaviour gives rise to emergence. Emergence is a phenomenon that generates patterns from the interactions of constituent parts that is greater or different than the summation of those parts. When local interaction parameters go beyond a critical threshold emergent properties may occur and new hierarchical global structures may emerge [39, 40]. The Schelling segregation model [41], Sakoda Checkerboard model [42], and Sugarscape [43] are classic examples of how emergent behaviour can result from individual preferences (local scale) for neighbour type. GIS are particularly important when exploring emergence because provide a source of information on neighbourhood interactions as seen in Fig. (**1**).

iv. The concept of emergence is linked to self-organization. This idea implies that some kind of system organization rises from within due to internal dynamics. In a self-organizing system the actions of individuals and local interactions are the source of system organization at a macro level. Epstein [33] and Von Bertalanffy [44] both argue that systems are more than a simple sum of the components, thus the need to analyse it as a whole due to this emergent patterns at larger scales, summarizing the idea that the large-scale patterns are the emerge from interactions at the local level. The dynamics of a system cannot be understood from decomposing system into its components, as apparently insignificant local interactions can lead to overall global different structures [39, 45].

v. The effect of any action can feedback onto itself or another action, directly or indirectly after several time steps. These feedback processes or so called loops can be positive, if amplifying or enhancing some effect, or negative if stabilizing or detracting an action. Feedback processes emerge from the relationships among system components which become more important than their own specific characteristics, building cycles that alter micro-level interactions, macro-level patterns, and the computational constructed environment, influencing the overall behaviour of the system [46]. Cilliers [47] refers to this term as recurrence.

vi. Information and representation of many components, entities, interactions and functions often cannot be precisely localized, being physically or virtually distributed on various sites, thus the distributed nature of complex systems. GIS allow to use real-world geographical information in ABM thus conferring geospatial accuracy to the underlying processes and dynamics in study.

vii. Complex systems are self-similar implying that patterns will repeat themselves at different spatial and temporal scales, *i.e.* there is a fractal geometry consisting of self-repeating patterns structures on all length scales.

viii. Systems have limited decomposability, which means that it is quite impossible to understand its dynamic structure by decomposing it into isolated components. Interaction among components and with the environment are crucial to grasp systems' self-organization and emergent properties. Thus, a large number of elements is usually necessary, but not sufficient.

ix. Each element in the system is unaware of the behaviour of the system as a whole, reacting only to information that is available to it locally. This issue is central to complex systems, because individual components don not have full awareness of global systems processes and dynamics. Therefore complexity emerges as a result of the patterns of interaction between the elements.

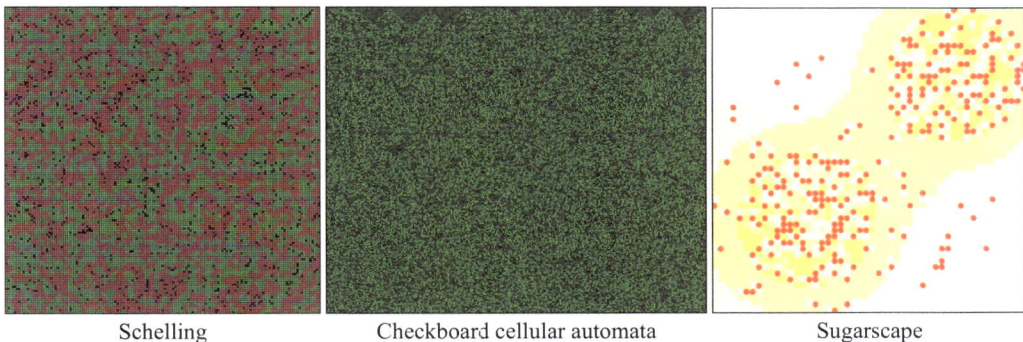

| | Schelling | Checkboard cellular automata | Sugarscape |

Fig. (1). ABM geographic information based environment in the study of emergence. From left to right: Schelling model, Checkboard cellular automata model, and Sugarscape.

Complex systems are composed of a myriad of heterogeneous objects called agents that act driven by specific goals and their interaction, amongst each other and with the environment, generating emergent local patterns that produce macro-scale phenomena often at higher levels than those at which such agents operate [48]. ABM provide a means to explore simulations of autonomous heterogeneous agents, in a designed spatial environment, allowing them to interact according to simple local rules that are conceived to mirror (in a simplistic way) real-world behaviours. It is important to keep in mind that ABM are abstract and simplified representation of reality and therefore can only do so much in transposing reality to "virtuality", explaining observed phenomena or foreseeing future realities.

Since, for some authors like Franklin and Graesser [49], the notion of agent is controversial, we hereby adopt Macal and North [50] structure of an agent-based model, which is comprised by three elements: 1. the agents, their attributes and behaviours; 2. agent relationships and interaction; 3. agents' environment.

Agents, Attributes and Behaviours

Most authors agree that, although there are multiple definitions of the term "agent", several attributes can be pointed out such as: heterogeneity, autonomy, capacity to process and exchange information, follow if-then rules, goal-driven, deductive code based units, with boundary and state. There are also core behaviours: mobility, interaction, adaptation, bounded rationality [7, 15, 18, 48 - 55]. These attributes and behaviours are hereby further discussed.

Agents are not aggregated into homogeneous pools. Rather, agents are heterogeneous possessing different attributes, behaviours and rules that may differ in multiple ways Agents are autonomous entities that are not subjected to the influence of external direction. They have the capacity for processing information and share it with other agents, though individual based interaction. An agent may be goal-driven and takes independent actions to reach its goals which can be described by simple if-then rules used to describe the theoretical assumptions of agent behaviour. ABM are mathematical and deductive computer programs, typically coded in a structured or object-oriented programming language.

One can easily determine agent boundary. Agents have attributes that allow to distinguish and recognize them. It is individually identifiable whether an attribute is or not part of an agent. The attributes influence agents state that represent the essential variables associated with its current situation, which can change over time and influence agent's behaviour according to a predetermined set of rules. State varies over time and according to the model objective, for instance in an epidemic model agents can be infected, susceptible, or resistant to virus [56]. The overall state of an ABM is the result of the aggregate state of its agents plus the state of the environment.

Behaviours can be theoretically framed in behavioural modelling frameworks, based on for instance heuristic like genetic algorithms, empirical experience, pattern recognition like machine learning, to name a few. In ABM, the Belief-Desire-Intent (BDI) cognitive architecture is one of the most used for defining intelligent and autonomous agents, endowed with processes of rational 'thought', allowing for the computational simulation of cognition and human behaviour [57].

By using BDI approach one can model agent's internal state through three types of mental processes: beliefs, desires, and intentions. Based on these, a control architecture is defined that enables the agent to select, in a deliberative way, its course of action. There are other alternative models such as: trans-theoretical model of change, health belief model, reflected theory of action, planned behaviour theory (among others). However these models share certain key characteristics: (a) the importance of knowledge of the world by the agent, (b) multiple levels of behaviour influence (c) behavioural changes as a process, (d) motivation *versus* intention, (e) intention *versus* action, and (f) behavioural change *versus* behaviour maintenance [58].

Mobility is another central feature of agents in ABM. It is particularly useful in spatially explicit models, composed of grid-based environment, n-dimensional lattice, or dynamic social networks. ABM can also be non-spatial. However, in most ABM the main notion is of local or location [43, 47]. Models in which agents roam through the simulation environment are frequent. Head, Rand and Wilensky [59] use ABM for traffic simulation where cars are the agents that follow a simple set of rules: decelerating if there is a car close ahead, and accelerating if there is not a car ahead. Another example is the path-finding ABM developed by Zhou [60] using the A-star algorithm to find the shortest path between origin and a randomly selected destination. In the ABM commuters move one node in a tick. When they reach the destination, they stay there for one time step and then move on to the next destination. This example demonstrates the interest in connecting agents and geographic data [15] through the coupling of GIS and ABM. In ABM, agents usually have spatial relationship between them and interact over an artificial environment. GIS typically contain multiple layers endowed with attributes (*e.g.* tourists, road network, or administrative boundaries) that can enhance environment representation capability when modelling. The stacking of multiple layers allow to model agents to move around a geographical environment characterized by vector data (line, point, polygon) or in a raster represented environment.

Typically, agents are social and interact with other agents (neighbours or non-neighbours) in the simulation environment. Following predetermined set of rules, agents have the ability to exchange information with other agents and with the environment. Agents do not need to be spatially linked to interact, taking from the example of the small world theory, or the concept of six degrees of separation [61], which argues that a person is only a couple of connections away from any other person in the world. The use of this approach is starting to be recurrent in social systems simulation, as many ABM are based on phenomena that requires information from online social networks, where the average degree separation of the agents' subtract network does not require direct contact between agents.

Agents have bounded rationality, which revises the assumption that individual decision-making is based on rational optimization [15]. Perfectly rational decisions are often not feasible because of two principles: i) bounded information access, stating that agents do not have access to system whole/global information; ii) and bounded computing power for agents do not have infinite analytical power [62]. Agents are also bounded by their cognitive abilities which will depend on the level of detail the developer put into coding its attributes and behaviour, and also the finite amount of time to make a decision [29, 47]. Therefore bounded rationality provides a more realistic individual decision-making taking into account time constraints, cognitive ability and incomplete information access. Agents make decisions to satisfy their goals, but usually not in an optimal way [63 - 65].

By having rules (or more abstract mechanisms) agents adapt their behaviours. Adaptation requires some kind of memory storage, which allows to access stored previous experiences and information shared with other agents and the environment, in order to learn and evolve, not necessarily in the most efficient way due to agents bounded rationality. Adaptation is a feedback process with which agents can constantly alter their state depending on their current state.

Relationships and Interactions: How and With Whom?

The main focus of ABM is on modelling interactions and relationships of agents, between agents and with the environment. There is typically an underlying topology (Fig. **2**) which define how and with whom agents are connected with, and the mechanisms of the dynamics of the interactions. Topology or connectedness usually include a network of nodes (agents) and links (relationships) describing information transfers [66].

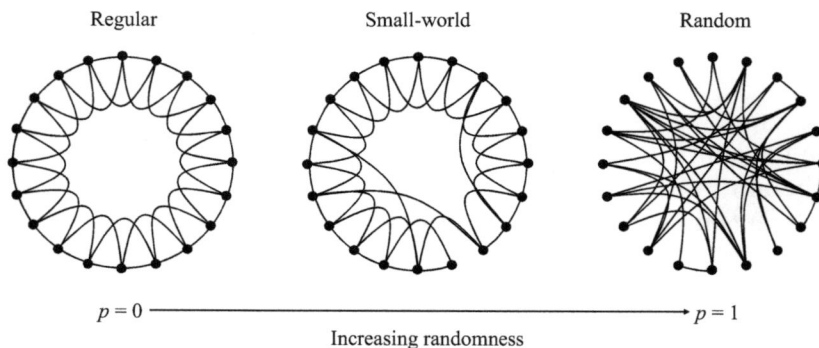

Fig. (2). Network topology according to Watts and Strogatz [66].

Agents have access to local information that is shared in direct or indirect ways, meaning that agents interact with other agents, but not directly with all the other agents. Agents obtain local information directly through interaction with neighbouring agents, using as substrate a network that defines maximum reach, and from its localized environment that can be grid, lattice or fractal based (*e.g.* with Moore's neighbourhood in a cellular automata grid). An agent's set of neighbours can change as a simulation proceeds and agents move through space, for instance as happens in the Follower ABM where agents attempt to connect with other agents, forming long chains according to a small set of simple rules [67].

Indirect interaction is also possible in ABM, in the sense that agents can interact with other non-neighbouring agents. To mimic real-world phenomena, agents can be inserted in social networks for instance relatives and friends, and online social networks. This allows for an agent to interact not only with its neighbours located close-by in geographical space, as well as neighbour agents located close-by in its social space as specified by the agent's social network. There are some ABM that use the small world theory to mimic online realities where agents can be connected to any other agent [61, 68].

A typical agent structure encompasses agent intrinsic characteristics, agent interactions with the environment, and agent interactions with other agents. In an ABM every characteristic associated with an agent is either an attribute or processes. Agent attributes can be static, not changing during the simulation like for example agents' designation, or dynamically changing as the model runs, *e.g.* state.

Agent processes consist mainly on behaviours that tie agents' goals with potential actions. ABM are composed of cognitive agents each possessing its own behaviour-model determining its interactions with other agents and the environment. When interacting agents can have active or reactive behaviours. Active agents exert independent influence in a simulation and can often be pro-active. Pro-activity is regarded as a goal-driven behaviour. Reactive agents are contrary from active ones. Their behaviour is usually defined by if-then rules and is resultant of an external trigger from other agents or the environment [69].

Simulation Environment

Agents' interaction is spatially bounded by the simulation environment, which defines the space where agents move. Agents may be spatially explicit, with precise location in a geometrical space that can be for example grid, lattice or fractal based, as shown in Fig. (**3**).

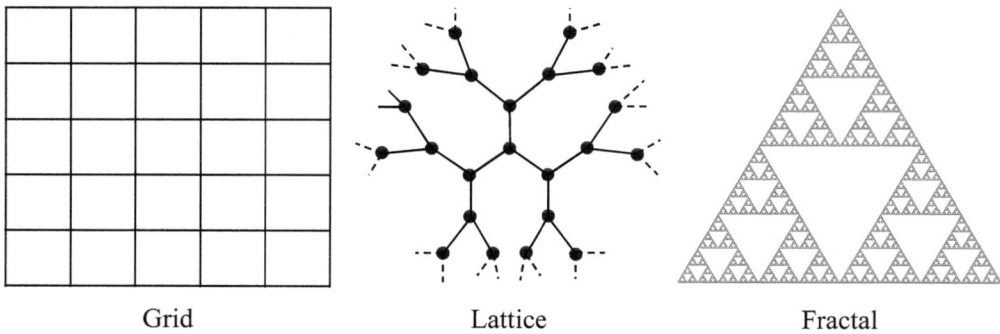

Grid	Lattice	Fractal

Fig. (3). Types of simulation environments.

Environment can also be static, remaining constant throughout simulations. Real-world environments are usually highly dynamic, subject to changes either caused by agent interaction with each other or with the environment, or subject to other processes independent from agent's behaviours. This dynamism contributes to emergence and feedback processes (earlier discussed).

Agent location within a simulation environment can be spatially explicit or implicit, meaning in the first case its position is known or relevant for the simulation, and in the second case is unknown or irrelevant. For example, in a pedestrian flow model agent location is essential [15, 50, 60].

The environment can provide information on the spatial location of an agent relative to other agents or it may provide a rich set of geographic information about the environment, as in a GIS, *e.g.* hydrography, altitude, accessibility. The majority of GM represent space as a matrix of discrete cells [70] providing valuable geographic detail but lacking geometric detail. This point is critical to models outcomes quality but is rarely referred in the literature [12, 71]. The capability to represent real world features in vector format, *i.e.* as points, lines and polygons, introduces geometry into the simulation process, allowing one to make more realistic models. The use of vector based GM is a step forward from the matrix structures used in early models [40], specially the use of polygons (*e.g.* administrative units) to constraint the agent interaction environment. Objects in GM can be understood as spatially fixed (*e.g.* beaches) or non-fixed (*e.g.* tourists). Both can have related transition rules which give them specific functionalities based on their spatial location and neighbourhood. Nevertheless, it is understood that areas usually do not change. Changes are expected to occur due to interactions between agent s and/or between agents and their environment [17] (see Fig. **4**).

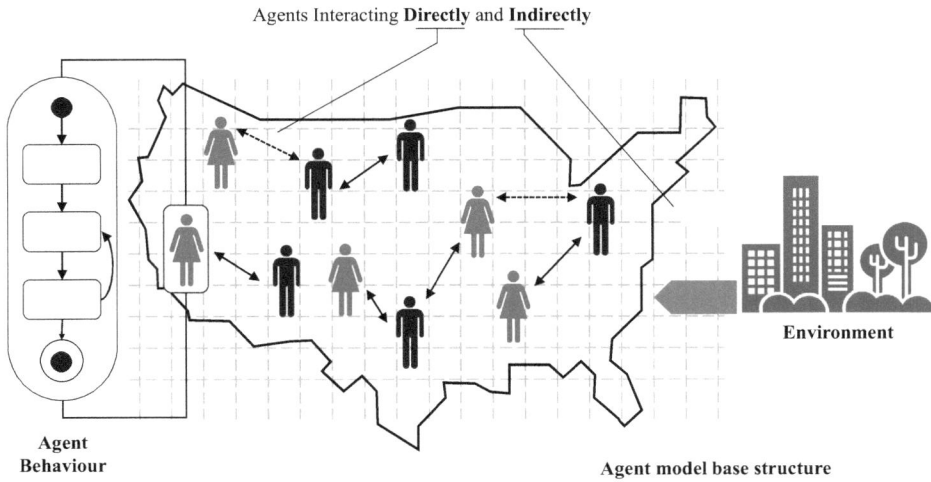

Fig. (4). Scheme of ABM structure.

Thus, an environment can be accessible when agents can obtain complete, updated information, or as happens in most real-world environments that are inaccessible, in the sense agents cannot access information isotropically.

ABM APPLICATION PURPOSE TYPOLOGY

The application purpose of an ABM determines the ways in which models can or should be used, regarding main aim and necessary data. When using an ABM, one has to make a commitment between the purpose of the model and its precision, that is, the type of data and knowledge that is required. Several authors propose ABM application purpose typologies and although the nomenclature may not be consentaneous there is a general agreement that ABM can be categorized in: theoretical or exploratory models, and empirical or predictive models [15, 46, 62, 72 - 74].

Castle and Crooks [15] categorization seems to be a quite clear reference, separating the utility of ABM into: explanatory or predictive. The explanatory modelling approach is proof of concept oriented aiming to explore theory and derive theoretical hypothesis. The main purpose is not to predict the future behaviour of a system, are usually built for illustrative purposes to provide a framework that is based on past information as an attempt to reproduce the system present situation [62], producing similar trends and patterns to those registered in real-world. These models can generate insights about theory and thinking on complex systems.

Predictive models are used for forecasting trends and scenarios, aiming to mimic real-world systems and to predict the future state of the system. These models usually use detailed empirical real-world statistical data. However, model developer should be aware that predictive models can be overly-fitted if parameterized with too much real-world data.

Johnson [46] also provides a good overview of ABM application purpose typology based on a comparison between application purpose (ranging from applied to theoretical) and representational scope (from general to detailed). As one moves forward from theoretical general models to applied detailed ones, the level of complexity and validation requirements increases. The author defines four general categories of ABM: proof of concept, hypothesis testing, pattern replication, and planning support. Proof of concept models are designed to represent the system in a highly generalized fashion, exploring theoretical hypothesis. Hypothesis testing models require a higher level more detailed representational scope (behaviours and environment), providing *in silico* simulation environment that represent realistic interactions for scenario testing [4].

ABM that attempt to replicate patterns provide a "social science laboratory" for developing a better understanding of pattern and process formation [7]. Have an applied purpose, although having quite general requirements for representational data. The fourth category of Johnson's ABM application purpose typology is that of planning support. It the most complex to apply and requires highly detailed and extensive data about agent behaviours and the simulation environment. It aims to model real-world interactions and scenarios to support planning decisions, such as for instance evaluating policies and action outcomes, and if possible generating new options.

Therefore, the choice of adopting an explanatory or predictive approach in an ABM is dependent on the purpose of the model and its precision. The availability of data and knowledge required for modelling also influences the choice of the approach.

ABM AND TOURISM RESEARCH

Understanding tourism has been recently recognized by the academic community as a complex problem. However there has been little empirical research into complexity and chaos theory in tourism research, and despite growing interest the first approaches are very recent [25, 75 - 78]. The relationship between ABM and complexity is reciprocally favourable: complexity theory provides the theoretical background and concepts for ABM, while advances in ABM provide tools to explore complex phenomena. Through this approach one can represent systems comprising components that (i) display non-linear relations, (ii) have thresholds, (iii) possess memory, (iv) are path dependent and (v) have learning and adaptation capabilities [7, 54], as previously discussed.

ABM can be thought of as a social virtual laboratory provide a tool to "think with" [79]. There are some studies that are more focused on grounding theoretical basis to frame complexity in tourism. In the turn of the century, seventeen years ago, Bob McKercher [25] already argued that tourism should be addressed as a chaotic, non-linear, non-deterministic system. The author pointed out that existing tourism models fail to fully explain the complex relationships between the systems components. Therefore an alternative model of tourism was proposed based on the principles of chaos theory. Studies from authors like Baggio [31, 80, 81] also follows the same approach presenting an overview of the complexity framework as a means to understand structures, characteristics, relationships, and explores the implications and contributions of the complexity literature on tourism systems.

Several authors adopted this theoretical viewpoint applying ABM to model tourism as a complex system. There are some applications of ABM to the study of tourism systems, varying from explanatory to predictive, but all of them quite recent. We hereby present some of them. For instance, Baggio and Baggio [82] developed an exploratory ABM to verify the relationships between the attractiveness of a tourist destination and tourist arrivals. The study concluded that in each time step agents act according to the defined rules in order to meet their goals, repeating the process until the selection of the best destination is true. The process agents go through to select the destination where their personal criteria are fulfilled, constitutes an adaptive behaviour.

One of the first applications of ABM coupled with GIS is the Recreation Behaviour Simulation (RBSim 2). RBSim 2 is a tool designed for simulating human recreation behaviour in in real world environments, allowing recreation managers and researchers to explore scenarios of changes in tourism systems [83].

Johnson and Sieber [46, 75, 84, 85] developed a pivotal study regarding the use of ABM as a potential tourism planning support system (PSS) in a Canadian province. This study is one of the firsts to integrate complexity theory in tourist system modelling. The authors provide a simulation model and platform – TourSim, which can be presented and used by planners as a PSS. The results of this research indicate that the use of an ABM-based PSS is strongest as a scenario development tool, providing an environment for formulating 'what if' scenarios, data analysis, and a tool for communicating results.

Other authors, such as Chao, Furuta and Kanno [86] developed an ABM combined with GIS which intended to understand the development process of Recreational Business Districts in tourism areas in East Asia. The ABM aims to provide a framework for supporting sustainable tourism development supporting policy makers and tourism bureaus assess different tourism policy scenarios and improve tourism services.

Another approach was adopted by Doscher *et al*. [87] by developing an ABM to model of tourism movement impacts and to allow for the evaluation of different scenarios (such as increases in petrol prices or variations in currency exchange rates) on the behaviours of tourists in New Zealand.

In the last years several developments have been carried out in ABM applied to tourism. An example is Pizzitutti, Mena and Walsh [88] that developed a touristic activity ABM to represent the touristic market, focusing on touristic offers, reservations, and touristic activities. The ABM is based on an individual-based representation of tourists' consumption preferences and touristic accommodation. Tourist agents are assigned with average characteristics of tourists to mimic the real world behaviours. The accommodation offers are is derived from data collected through field surveys. The model simulates three scenarios in order to study how emergent patterns in the touristic market can be affected by changes in the environment.

Also Boavida-Portugal, Ferreira and Rocha [68] developed an ABM to improve knowledge in the tourist decision-making process in selecting a destination to vacation. The ABM operates taking into account the tourists' social network and individual level, in order to represent and explore the basic theoretical principles of tourist decision-making behaviour. The authors design two simulations to explore how changes in the awareness for a destination and in tourist individual preferences change the destination selection. It is a pioneering study in the application of ABM to tourist decision-making process exploring individual subjective motivations and desires.

Some authors have applied ABM to model climate change impacts in tourism systems. Pons *et al*. [89, 90] provide an ABM planning support tool to consider the vulnerability of ski resorts in the Pyrenees region and help stakeholders design appropriate sustainable adaptation strategies to future climate variability. In 2013, Balbi *et al*. [91] use ABM to assess alternative strategies of adaptation to a changing climate and tourism demand in an alpine tourism destination.

Recently Nicholls, Amelung and Student [92] produced a paper aimed to a broader tourism audience, exploring tourism as a phenomenon to be subjected to ABM and the power and benefits of ABM as an alternative scientific mechanism in tourism studies. The authors also summarize the few existing applications of ABM in tourism focusing on the potential applications in tourism planning, development, marketing and management. This paper constitutes a good starting point for authors interested in developing tourism ABM applications.

POTENTIAL OF COUPLING ABM AND GIS IN TOURISM RESEARCH

Tourism is a multiscalar phenomenon, with several agents, that suffer influences, and produce impacts at different levels. In ABM agents often have some sort of spatial relationship to each other and can be situated in an environment. These agents act in, influence and are influenced by natural and human environments. Often there are feedback processes where the environment influences the agent which afterwards influences the environment. The integration of geographical information, in multiple layers (*e.g.* a housing layer, a road network layer, a population layer) each layer made up of a series of features (*e.g.* points or polygons) can provide valuable information regarding the agents geographical position in the environment and other agents, as well as from the environment in which they act. This ability to include different features and attributes of different layers from a GIS, allows for a more accurate representation of the tourism system when modelling.

Several authors [71, 93, 94] have argued that GIS is not a modelling platform and that it is not GIS design propose, lacking in performance particularly when handling with large datasets and/or number of iterations. The models implemented as direct extensions of GIS generally make two assumptions: all operations required by the template are available in the GIS, and GIS has computational performance enough to deal with the implementation of the model. Therefore, one of GIS advantages – its broader applicability – becomes a drawback and limits its applicability in modelling and simulation.

A solution has been the integration (also referred to as coupling or embedding) of GIS with ABM and/or other modelling tools, aiming to achieve a fully operational GM [95]. This integration has been the subject of various classifications defined by different terminology by different authors [94, 96 - 98], for example the use of the term integration, coupling or embedding. This calls for the necessity of clarification of concepts, aiming to identify a suitable scheme for developing GM. When both GIS and ABM already exist or the cost of programming one system functionality into another is too high, integration is a solution [99]. Hence, integration can be generally defined as the connection of two standalone applications through data transfer. Most authors classify GIS-ABM integration in three types which can differ in terminology, *i.e.* some authors use week, moderate, and strong integration [15, 96]; while others use loose, moderate and tight coupling. We hereby adopt the following categorization into:

- Loose coupling, usually involves unbound and independent operation of functions within each system with data exchanged between systems in the form of files. For example GIS can provide the base information for simulation environment, for example vector information containing location of tourism accommodation establishments. GIS information can also be used to prepare the inputs that are latter on incorporated in the ABM, for instance a raster file with a continuous grid of tourism hotspots resulting from a hotspot analysis in GIS software.
- Moderate coupling, implies remote procedure calls and shared database access between GIS and ABM systems, reducing the procedures execution speed and creates difficulty in computing processes in both systems ate the same time.
- Tight or close coupling, integrates simultaneously operations of systems allowing direct inter-system communication during programme execution that can be run on one or more networked computers.

In tourism research, the use of fully GM is still in embryonic phase. The few studies that use an ABM coupled with GIS approach (discussed in the previous section) follow a loose coupling strategy. Nicholls, Amelung and Student [92] point out an agenda for future development of ABM applications in tourism research and some of them, in our view, could benefit from loose coupling with GIS. Namely, for management and modelling of visitor flows ABM offers potential for crowd management and tourism-related risk assessment, especially at transportation hubs, major attractions, and large events. For this purpose geographic information is useful by providing environment characteristics, for instance road network, emergency exits, so that agents can move around a real world information fed simulation that will produce outputs like emergency procedures or plans.

Tourism is highly dependent and subject to climate conditions, which in part justifies the growing interest authors have shown in developing ABM to explore climate change impacts on tourism [89]. Some studies have incorporated GIS information to specify the simulation environment such as the location of sky resorts and, using a moderate coupling approach to produce in GIS layers that will be input parameters in the ABM computational functioning, such as *e.g.* vulnerable areas, energy/transport costs, demographic maps.

The setting up of the simulation environment with GIS can also bring other relevant advantages picking from cellular automata and neighbourhood analysis that have been more used in GIS in ABM environment. If interactions between agents is one of the core issues in ABM this relations usually have some substrate topology or network that can be developed in GIS and integrated in ABM with a moderate coupling or following a loose coupling integration be derived in ABM using GIS inputs For example Slumulation ABM explores how slums emerge and evolve using a Moore's neighbourhood grid in which agents act and are connected to each other [100].

There is potential for integrating other novel geographic data collection methods such as global positioning systems [101, 102], or georeferenced data from tourism related online social networks such as photos posted on *e.g.* Flickr or Instagram [103] or Angelina's Jolie georeferenced tattoo [104], sequence alignment [105], and automated web harvesting [106]. Also the opportunity of Big data will impose in a brief future challenges in integrating ABM-GIS-Big data from tourism related sources. Most of these methods are still in embryonic stage as far as collection and analysis as well as in ABM integration but the authors think this type of data will open way for great developments in data collection that can feed individual-based simulations.

In this chapter we present a theoretical ABM that has the purpose to add to the understanding of the tourist decision-making process, through controlled computational simulations, using different sources of data (geographical and statistical), and based on theoretical simplified assumptions.

TOURIST DECISION-MAKING ABM FRAMEWORK

As in any model, when developing an ABM it is assumed that simplifications of real-world problems and incomplete knowledge and data about phenomena is an unsurpassable issue. There is a model-building process in ABM that can be represented in several stages: (i) the conceptualization of the system; (ii) the definition of variables, parameters and data to incorporate in the model; (iii) setting if–then rules which replicate behaviours; (iv) coding and application of the model; v) parameterization, calibration and validation and vi) finally, scenario testing.

The ABM framework presented hereby is based on the work developed by Boavida-Portugal, Ferreira and Rocha [68] which relies on several theoretical assumptions of tourist behaviours. In fact, in order to understand where and why tourists' vacation requires knowledge on tourists' behaviour, which is determined by (i) push factors, that are factors internal to the individual; and (ii) pull factors which are external to the individual [107, 108]. According to Goossens [109] tourists travel to a destination if the benefits offered by the destination are perceived to satisfy their needs, within certain constrains such as time and money. The tourist considers alternative possible destinations against a list of criteria crucial to satisfy individual needs. For a destination to be considered as an alternative, tourist need to have some kind of information about it, that can be based on previous individual experiences the tourist had in that same destination or shared by its social network. The more an individual is satisfied with the destination of choice the higher it will be placed on the list of alternatives. Thus, tourists learn from experience and show destination loyalty by coming back and recommending the destination to others in its social network [110, 111].

Taking this theoretical approach, the ABM developed intends to reproduce tourist behaviour in a simple way. In order to model individual behaviour influence in the destinations choice, several parameters are incorporated and explored. The tourist decision ABM considers (i) individual motivations; (ii) rationality based on if–then rules; (iii) human emotions and satisfaction and (iv) the influence of the social network [18]. Lastly, the destination chosen is the one that best meets the tourist motivation [87, 112 - 114].

Several authors have proposed different methodological structure for model building. The ABM developed is based on the approaches from Sargent [115, 116], Rykiel [117], and Gilbert and Troitzsch [118]. The first step in the ABM development is abstracting agent behaviour from the target system. Model parameters are selected to represent realistically complex phenomena and can be for instance random, derived from statistical distributions, or according to heuristic decision trees [47]. The theoretical assumption abstractions are embedded in the model parameters, which are adjusted to real-world dynamics through the process of calibration. Calibration is the estimation and adjustment of model parameter values or thresholds, that are otherwise unknown [117]. The model is then run and the results of the simulation are subjected to the process of validation, which intends to demonstrate if the range of accuracy intended is consistent with the application purpose of the model. The model is valid if considered robust in supporting scenario testing. If it is not, the process of ABM building starts again with the alteration of theoretical assumptions and parameter thresholds.

CASE STUDY: TOURIST DESTINATION DECISION-MAKING IN PORTUGAL

Portugal is a country where tourism is an important economic activity that according to the World Travel and Tourism Council 2015 statistical data directly contributes to 6.0% of the country's gross domestic product in 2014 and 15.7% total contribution). Even more relevant is the contribution of this activity to employment: in 2014, were estimated 7.4% direct jobs and about 18.4% of total employment. The weight and the dynamic growth of this activity have led to processes of changing land use and functions of many spaces and places, particularly along the coast, where the demand associated with the sun and sea tourism has been the main engine of this transformation (European Environment Agency, 2006).

The ABM presented models the decision context of tourists coming to Portugal. The spatial analysis unit are based on Portuguese NUTS II, a geocode European Union standard for referencing the subdivisions of countries for statistical purposes (Nomenclature of Territorial Units for Statistics). However, the model is not spatially explicit yet, and does not incorporate geographical datasets nor parameters that reflect spatiality (like distance and cost to travel). It incorporates a substrate geographical network that allows to evaluate the agents' neighbourhood. In the future work section will be discussed how the authors are working on coupling a CA land-use change model into the ABM.

Setting the Agents: Tourists and Destinations

This ABM represents tourism as the relationships between the destinations and tourists (supply and demand), and tourists among themselves. Thus, there are two types of agents in the ABM, each with different attributes: tourists and destinations. Tourists (T) are mobile agents and have distinctive profiles that are related to their personal preferences. Each tourist has a personal motivation list (mot_T) with four key elements related with the activities tourists want to engage in: sun and sea; city break; food, wine and cultural touring; nature and active based (such as hiking, golf, surf, nautical). These elements and the tourist profiles were set up based on the Tourism National Strategic Plan (PENT) and Tourism 2020 National Strategy (T2020). For each element on the motivation list, the tourist has a value that corresponds to the drive for these activities in a destination; for example, if a tourist wishes city break type of vacation will have a high value in the city break parameter. The setup function generates five different kinds of tourist agent profiles (P): sun and sea seeker ($P1$); city break seeker ($P2$); food, wine and cultural touring seeker ($P3$); nature and active based ($P4$).

Destinations (D) are non-mobile agents and serve as a non-spatial environment for the simulation. There are seven destinations corresponding to Portugal NUTS II: Norte, Centro, Lisboa, Alentejo, Algarve, Açores, Madeira. Each destination has an attraction list (A_D) defining the most attractive resources of the destination: sun and sea; city break; food, wine and cultural touring; nature and active based. The values for the attraction are based on the main strategic tourism market segments pointed out by PENT and T2020 and by empirical knowledge of the regions' tourism characteristics.

Destinations have a specific tourist weight (weight T_D) which defines the maximum threshold of tourists for each destination and time step in order to make the vacation satisfactory. It is defined by a function of the ratio between the yearly total number of tourist for Portugal and the yearly total number of tourists for each destination.

ABM Parameterization

The outline of the tourist decision-making process developed in the ABM is shown in Fig. (**5**). At each time step, the tourist decision-making process in the selection of a destination depends on two main factors: the individual-level (I) and the social influence (SOC). The individual-level parameters refer to the personal characteristics of tourists and include the compatibility (C) between tourists' motivation list (mot_T) and destination attractiveness (A_D) and the individual satisfaction (S) of the last experience the tourist had in a specific destination.

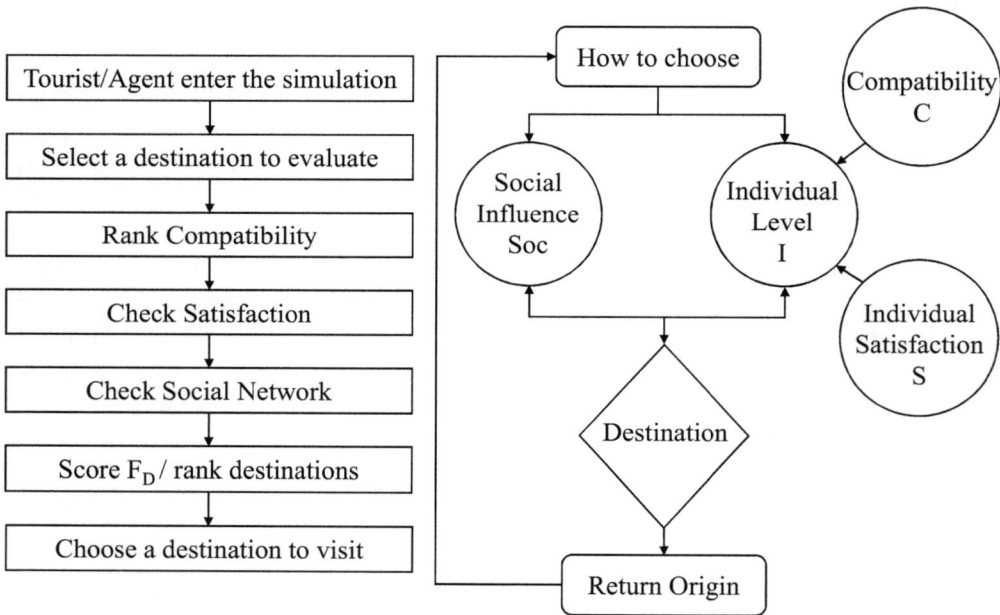

Fig. (5). Tourist decision making process adapted from [68].

Tourist agents are endowed with rationality and their choices with simple if–then rules, producing simulations that resemble real-world patterns. There are also abstract parameters that rely on abstract human characteristics, such as the social network influence and satisfaction of the vacation.

To match tourists' motivation with destination attractiveness, the model incorporates a compatibility parameter (*C*), computed in the individual level. Tourists with different profiles want to vacation to destinations that match their motivations. For instance, a tourist seeking city break-related activities is more likely to want to vacation in a city break type of destination. Thus the compatibility between the tourists' motivation list (*mot$_T$*) and the characteristics of the destination depicted in the attraction list (*A$_D$*) is fundamental in the ABM. The compatibility parameter is computed as

$$C_D = \sum_{i=1}^{n} mot_T - A_D \ .$$ (1)

When a tourist visits a destination, its satisfaction score is updated based on how many other tourists there are in the same destination at the same time, designated as occupancy rate. This parameter is based on the assumption that if the destination is close to the maximum number of tourists it usually receives (based on the specific tourist weight, the destination is crowded and the tourist is less likely to enjoy the vacation experience as much. Thus, the level of satisfaction decreases and tourists are less likely to revisit. The occupancy threshold varies between 0 and 1, computing a ratio between actual tourists and maximum occupancy for each destination; for example, if the value is 1, there is enough space for all tourists (100%), but if the value is 0.5, only 50% of tourists will have space and thus a satisfactory experience. There is also a satisfaction weighing parameter (weight *D*) with a threshold between 0 and 1, which defines the weight of the last and current vacation; for example, if the value is 0.5, the satisfaction of the last visit and the satisfaction of the current visit have the same weight in the computation for the satisfaction parameter. In order to make the satisfaction parameter more realistic, based on the principle that the vacation satisfaction depends on other variables rather than the crowdedness of a destination, some randomness was introduced in this parameter, with a random-float parameter that varies between 0 and 2.

The social influence is the second factor in the ABM and was incorporated in order to test the formation of networks that result in the "small world" phenomenon. The small world theory [61] is based on the idea that a person is only a couple of connections away from any other person in the world. Similarly, tourists are inserted in their own social network (family, friends or co-workers) with whom they share and discuss previous vacation experiences, usually giving an overall score to the destinations (*i.e.* individual satisfaction).

To replicate the small world phenomenon in the ABM was used a geographical network as substrate based on neighbourhood, with an average degree of four, *i.e.* where tourists are connected with their four closest neighbours. Tourists can also be randomly connected with other non-neighbour tourists through an adjustable rewiring probability incorporated in the model. The rewiring probability is a parameter of the model with a threshold between 0 and 1; that is, if the value is 0.5, there is a 50% probability to be connected to another tourist who is not one of the four closest neighbours. This parameter aims to represent social interactions in a more realistic way, in the sense that tourists can share information by other means of communication, such as in an online social network.

Thus, a subtract network based on neighbourhood is created incorporating information sharing among tourists about previous vacation experiences. Tourists share and inform connected neighbours about their individual satisfaction score (I_D) for each destination they already visited. The social influence factor (SOC_D) is calculated as a mean of connected neighbours' satisfaction score for previous vacations at a destination; that is, if my four connected neighbours share values for satisfaction score for destination x of $\{1; 0.3; 0.5; 0.7\}$, the satisfaction score is computed as 0.625. The influence that the social network has on tourist destination choice is a model parameter with a threshold between 0 and 1. There is also a social network weighing parameter that allows the modeller to weight the influence the social network has on the final score F_D for the choice of the destination.

The modeller can change parameter thresholds and run different simulations. At each time step the tourist agent computes a final score F_D for each destination using

$$F_D = SOC_D \cdot I_D .$$

(2)

This way the agent ranks the destinations according to the F_D value and chooses to vacation in the one with the highest score.

ABM Calibration and Validation

To test the robustness of the model, the parameter values were tested. Several simulation were carried out (using sensitivity analysis) to confirm the impact each parameter had on simulation results. Then, the parameter values were calibrated until the results of the simulation significantly reproduce the real total number of tourists. The simulation parameter thresholds found to best mirror the total number of tourists/yearly for each NUTS II were: (all parameters on) satisfaction weight 0.60; social network rewiring probability 0.50 and social network weight

0.40, and the occupancy rate was set at 0.55. The process of validation using regression analysis allows to evaluate if the model produced significant results. The results show that there is a significant positive correlation between the real total number of tourists and the simulation results, as can be seen by the R^2 for each destination in Fig. (**6**).

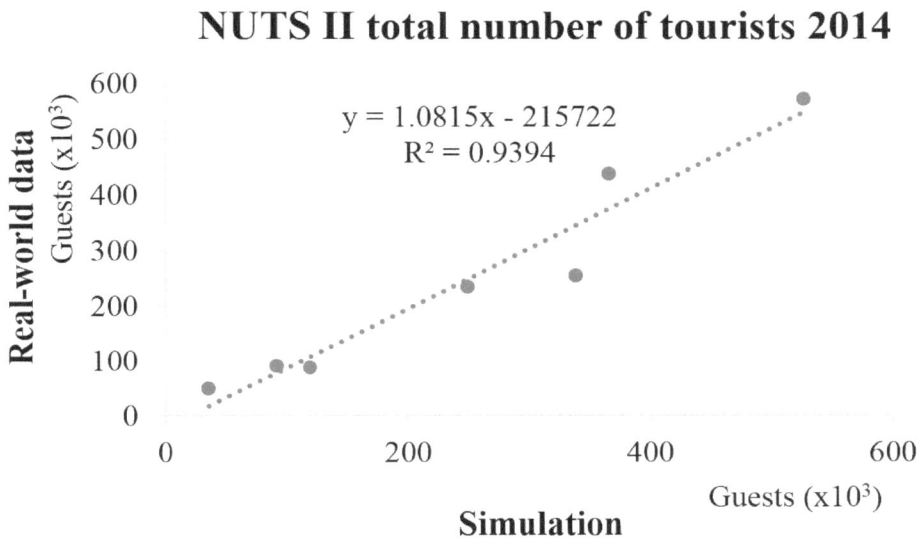

NUTS II total number of tourists 2014

$$y = 1.0815x - 215722$$
$$R^2 = 0.9394$$

Fig. (6). Regression analysis for the real-world total number of tourists and simulated results for 2014.

RESULTS AND DISCUSSION

Different simulations were performed in order to explore tourist decision making process. The setup process assigns the values for the tourist motivation and destination attractiveness. Then the user defines the ABM parameters thresholds and the simulation runs 365 time steps (one year period). Each time step a normalized value of 470000 tourists enter the simulation environment in search for the most suitable destination following the discussed theoretical framework and according to parameter values. Tourist profiles were set based on PENT and T2020 as: 25% sun and sea; 40% city break; 15% food, wine and cultural touring; 20% nature and active based.

Several simulations were performed in order to reach the conditions found to best mirror the real-world system under study, *i.e.* the total number of tourists/yearly for each Portuguese NUTS II. We hereby present only a few simulation results, expressed in Fig. (**7**). SIM1 is the situation found more suitable for this goal, with

R^2 0.94 as seen on Fig. (6). All parameters were on, and the threshold values used for each parameter were: satisfaction weight 0.60; social network rewiring probability 0.50; social network weight 0.40, and the occupancy rate was set at 0.55. There is indication that the number of previous satisfactory trips to a destination increases the probability of returning and fulfilling personal motivations. Also with the increase of awareness, or connectedness with the social network (defined by the rewiring probability), tourists will take more informed decisions and they are more likely to choose a destination that meets the expectations, which leads to a satisfactory vacation.

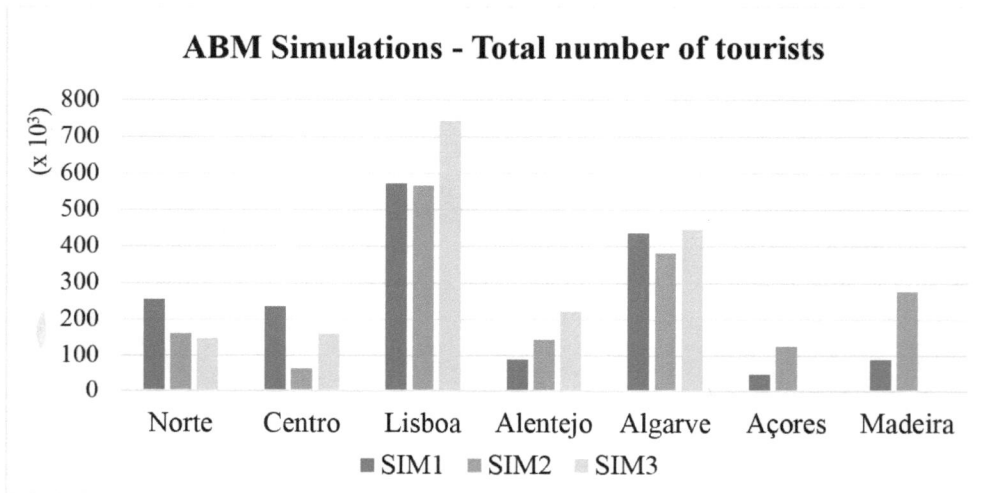

Fig. (7). Total number of tourists in simulations performed.

In SIM2 intended to test the satisfaction parameter importance in the results, thus the satisfaction parameter was off. The remaining parameters were kept equal as SIM1. The distribution of tourists in the alternative destinations produced rather different results, with R^2 0.53 when compared to real-world distribution, showing that previous vacation experience have significant impact in the choice of destination. Whenever the satisfaction perception of vacation experience in a destination is not a criteria in the choice of the destination, tourists go to other destinations where their compatibility parameter is met regardless of the experience evaluation.

In order to test social network influence SIM3 was carried out with social network parameter off, but maintaining SIM1 other parameters values. There is significant positive correlation of R^2 0.73 between simulated and real-world destination tourist distribution values. That is to say that with no level of awareness of destination satisfaction or recommendation from the social network, tourists will take less informed decisions and they are less likely to choose a destination that meets the expectations, leading to a less satisfactory vacation. When tourists share information about previous experiences in the destination, such as in SIM1, they are more likely to adapt their behaviour and travel to destinations that can be more suited to their expectations.

The presented scenarios tested, however based on theoretical hypothesis, point out that some relevant aspects such as the previous knowledge of destination, through sharing with social network, can influence the vacation experience. The results also provide information about the individual preferences of the tourists, namely that compatibility and satisfaction level from previous vacation experiences introduce different behaviour patterns. In fact, the model shows that tourists are pushed by their own internal forces such as motivation and satisfaction. These behaviour patterns have been discussed by authors, such as Baggio [31] and Mill and Morrison [119], and are tested in the presented developed exploratory model.

Thus, we observe the way in which changes in the model parameters generate different patterns of visitation. These emergent patterns are driven by the interactions between tourists and with destinations, and by the feedback processes produced among them, which also infer with the destination decision-making process.

Future Work

The future (and on-going) work in the tourist decision making ABM is to introduce interactions with the environment and the adaptation of the tourism system supply to demand trends. A CA land-use change model was developed (for one Portuguese NUTS II) in order to derive a 2020 scenario for tourism related land-use change (see Fig. **8**).

Using a Cellular Automata/Markov chain land-cover change prediction method several parameters where computed to reach the most desirable and suitable cells for tourism development, such as: proximity to most attractive touristic spots, proximity to roads and infrastructure, distance to urban settlements, distance to the coast, distance to existing tourism accommodation establishments, land-use cover change from the period of 1995-2007-2010, elevation, land-use types, ecological restrictions, and existing land-use areas defined by planning instruments as touristic prone.

The CA output will be coupled with ABM through a loose coupling approach. Tourist agents will then interact with the CA built environment producing changes as the demand for one destination goes up and accommodation capacity is reached new tourism accommodation establishments will emerge in the most suitable cells embedded in the CA.

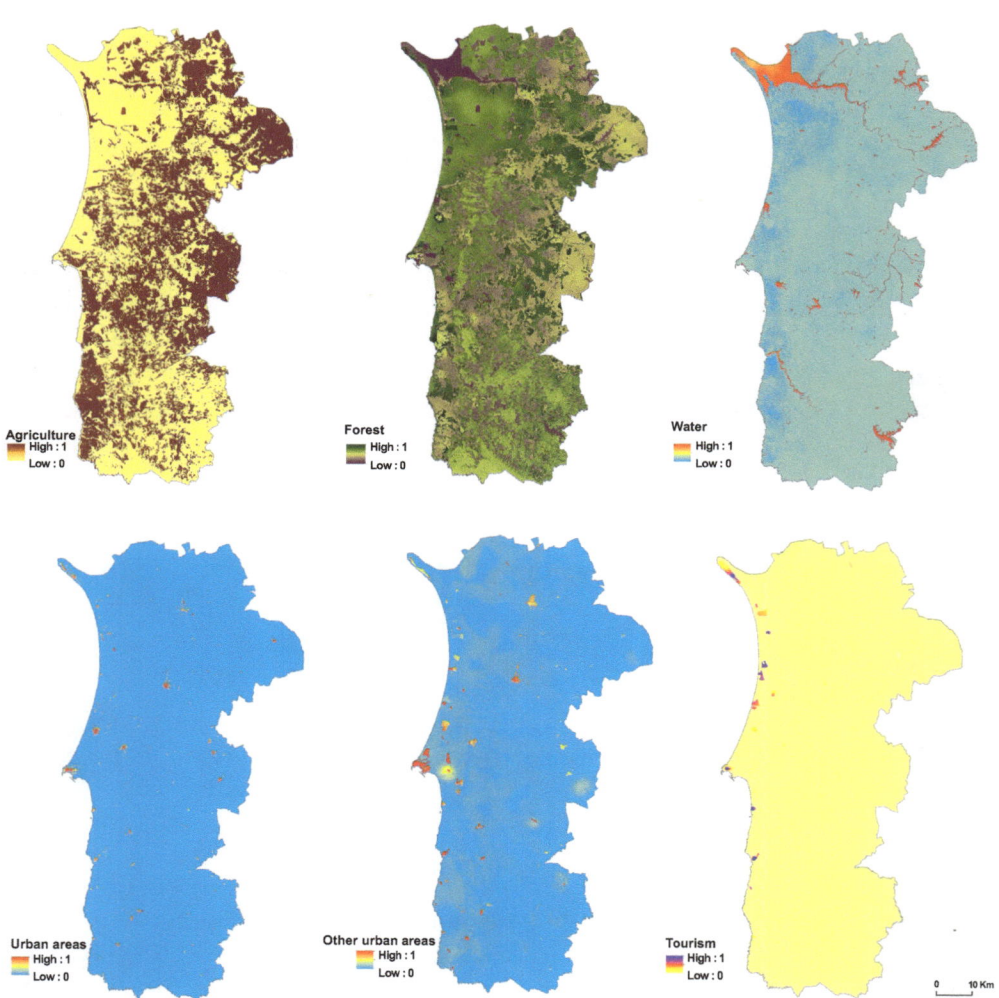

Fig. (8). Example of input and output data of the Cellular Automata/Markov chain land-cover change prediction method.

CONCLUDING REMARKS

In this chapter, the aim was to develop an exploratory ABM, even with all the simplifications introduced, to test tourist individual decision-making process in selecting a vacation destination. Several simulations were performed, by changing model parameters, to present results that reasonably reproduce real-world tourism system patterns. Even though the model presented is a basic representation with strongly simplified assumptions of the system processes, it aims to contribute to the understanding of how tourism demand patterns are formed, in order to explore basic theoretical principles of tourist decision-making behaviour [18, 111, 120].

However, there are several challenges in developing and implementing a tourism applied ABM, just to name a few examples: deepening the understanding of the system dynamics; lack of available micro-datasets; model validation process; and the lack of adoption of a standard model documentation procedure so that transparency and model replication by other researchers can be boosted. Also the gap existing between academic research and tourism practitioners makes the communication and acceptance of ABM results more difficult. Notwithstanding, ABM can be a useful tool for researchers and stakeholders in producing new insights about system dynamics, and providing a social laboratory to experiment "what-if" scenarios or more empirical models that can replicate real-life problems.

Other key point is the visualization of GM outcomes. Here the integration of ABM and GIS is fundamental. Often visualization is limited to a set of static frames of model results at best presented as an animated sequence. ABM allow to understand the behaviour and dynamics of the system instead of just look at an endless collection of statistic data. Nonetheless, such as in GIS, often the outputs visualization of GM are graphically poor [121], which has a detrimental effect on the transmission of information in particular to decision makers. There is the conception that models producing spatial explicit outcomes susceptible of being mapped should also be visually appealing [122]. However, it is also important to keep in mind that one of ABM main tasks is to allow a quick and clear preview of simulated scenarios [50, 74, 121]. Therefore GM tools, contemplating the coupling of ABM-GIS, could also benefit from some kind of integration with data visualization (*i.e.* datavis) tools.

The linkage to GIS is not a panacea for all the problems as is not recognize as a good modelling platform [94, 99], especially in terms of dynamic simulation. This presents two difficulties to GIS: the problem with handling time [123] and to represent continued variations. Still, the advantages of linking GIS with ABM overcome the remaining problems, especially when spatial and temporal analysis are necessary.

CONSENT FOR PUBLICATION

Not applicable.

CONFLICT OF INTEREST

The author (editor) declares no conflict of interest, financial or otherwise.

ACKNOWLEDGEMENTS

This work was supported by FCT – Fundação para a Ciência e a Tecnologia [grant SFRH/BD/75984/2011].

REFERENCES

[1]   D.G. Pearce, "Towards a geography of tourism", *Ann. Tour. Res.,* vol. 6, pp. 245-272, 1979.
[http://dx.doi.org/10.1016/0160-7383(79)90101-4]

[2]   L.S. Mitchell, and P.E. Murphy, "Geography and tourism", *Ann. Tour. Res.,* vol. 18, pp. 57-70, 1991.
[http://dx.doi.org/10.1016/0160-7383(91)90039-E]

[3]   A. Lew, "Defining a geography of tourism", *Tour. Geogr.,* vol. 3, pp. 105-114, 2001.
[http://dx.doi.org/10.1080/14616680010008739]

[4]   I. Benenson, and P.M. Torrens, "Geosimulation: object-based modeling of urban phenomena", *Comput. Environ. Urban Syst.,* vol. 28, pp. 1-8, 2004.
[http://dx.doi.org/10.1016/S0198-9715(02)00067-4]

[5]   R. Najlis, and M.J. North, "Repast for GIS", *Proceedings of Agent 2004: Social Dynamics: Interaction, Reflexivity and Emergence,* 2004

[6]   D.G. Brown, *Agent-Based Models.,* H. Gist, Ed., Greenwood Publishing Group: Westport, CN, 2006, pp. 7-13.

[7]   E. Bonabeau, "Agent-based modeling: methods and techniques for simulating human systems", *Proceedings of the National Academy of Sciences of the United States of America,* 2002, pp. 7280-7287.
[http://dx.doi.org/10.1073/pnas.082080899]

[8]   D.G. Brown, R. Riolo, and D.T. Robinson, "Spatial process and data models: Toward integration of agent-based models and GIS", *J. Geogr. Syst.,* vol. 7, pp. 25-47, 2005.
[http://dx.doi.org/10.1007/s10109-005-0148-5]

[9]   D.C. Parker, *Integration of geographic information systems and agent-based models of land use: prospects and challenges.* GIS, Spat Anal Model, 2005, pp. 403-422.

[10]   M. Haklay, D. O'Sullivan, and M. Thurstain-Goodwin, "' So go downtown': simulating pedestrian movement in town centres", *Environ. Plann. B,* vol. 28, pp. 343-359, 2001.
[http://dx.doi.org/10.1068/b2758t]

[11]   M. Batty, J. Desyllas, and E. Duxbury, "Safety in numbers? Modelling crowds and designing control for the Notting Hill Carnival", *Urban Stud.,* vol. 40, pp. 1573-1590, 2003.
[http://dx.doi.org/10.1080/0042098032000094432]

[12]   I. Benenson, I. Omer, and E. Hatna, "Entity-based modeling of urban residential dynamics: the case of Yaffo, Tel Aviv", *Environ. Plann. B,* vol. 29, pp. 491-512, 2002.
[http://dx.doi.org/10.1068/b1287]

[13]   P.M. Torrens, "Simulating Sprawl", *Ann. Assoc. Am. Geogr.,* vol. 96, pp. 248-275, 2006.
[http://dx.doi.org/10.1111/j.1467-8306.2006.00477.x]

[14]   L. Smith, R. Beckman, and D. Anson, "TRANSIMS: TRansportation ANalysis and SIMulation System", *Fifth National Conference on Transportation Planning Methods Applications,* vol. II, 1995, p. 10.

[15] C. Castle, and A. Crooks, *Principles and concepts of agent-based modelling for developing geospatial simulations. Paper 110,* 2006.

[16] A. Crooks, "Exploring Cities Using Agent-Based Models and GIS", *Proc Agent 2006 Conf Soc Agents Results Prospect,* 2006, pp. 125-132.

[17] D. O'Sullivan, and M. Haklay, "Agent-based models and individualism: Is the world agent-based?", *Environ. Plann. A,* vol. 32, pp. 1409-1425, 2000.
[http://dx.doi.org/10.1068/a32140]

[18] W.G. Kennedy, Modelling human behaviour in agent-based models.*Agent-based models of geographical systems,* A.J. Heppenstall, A.T. Crooks, L.M. See, M. Batty, Eds., Springer, 2012.
[http://dx.doi.org/10.1007/978-90-481-8927-4_9]

[19] M. Gell-Mann, Complex adaptive systems, *Santa Fe Institute Studies in the sciences of complexity proceeding.* Addison-Wesley Publishing Co., 1995, p. 11.

[20] U. Wilensky, *NetLogo Flocking model.* http://ccl.northwestern.edu/netlogo/- models/Flocking (1998).

[21] C.W. Reynolds, "Flocks, herds and schools: a Graphics, distributed behavioral model", *Comput. Graph.,* vol. 21, pp. 25-34, 1987. [ACM].
[http://dx.doi.org/10.1145/37402.37406]

[22] U. Wilensky, *NetLogo Wolf Sheep Predation,* 1997. http://ccl.northwestern.edu/netlogo/ models/WolfSheepPredation

[23] KJ Mock, and JW Testa, "An agent-based model of predator-prey relationships between transient killer whales and other marine mammals", *Univ Alaska Anchorage, Anchorage, AK, Tech Rep.*

[24] C. Cioffi-Revilla, "Computational social science", *Wiley Interdiscip. Rev. Comput. Stat.,* vol. 2, pp. 259-271, 2010.
[http://dx.doi.org/10.1002/wics.95]

[25] B. McKercher, "A chaos approach to tourism", *Tour. Manage.,* vol. 20, pp. 425-434, 1999.
[http://dx.doi.org/10.1016/S0261-5177(99)00008-4]

[26] E. Lorenz, The Butterfly Effect.*The Chaos Avant-Garde: Memories of the Early Days of Chaos Theory.,* R. Abraham, Y. Ueda, Eds., World Scientific Publishing: Singapore, pp. 91-94.

[27] V Grimm, E Revilla, and U Berger, Pattern-oriented modeling of agent-based complex systems: lessons from ecology, *Science,* vol. 310, pp. 987-991, 2005.

[28] R.J. Eidelson, "Complex adaptive systems in the behavioral and social sciences", *Rev. Gen. Psychol.,* vol. 1, pp. 42-71, 1997.
[http://dx.doi.org/10.1037/1089-2680.1.1.42]

[29] F. Reitsma, "A response to simplifying complexity", *Geoforum,* vol. 34, pp. 13-16, 2003.
[http://dx.doi.org/10.1016/S0016-7185(02)00014-3]

[30] P. Cilliers, and D. Spurrett, "Complexity and post-modernism: Understanding complex systems", *S. Afr. J. Philos.,* vol. 18, pp. 258-274, 1999.
[http://dx.doi.org/10.1080/02580136.1999.10878187]

[31] R. Baggio, "Symptoms of complexity in a tourism system", *Tour. Anal.,* vol. 13, pp. 1-20, 2008.
[http://dx.doi.org/10.3727/108354208784548797]

[32] R.K. Sawyer, *Social Emergence: Societies as Complex Systems.* Cambridge University Press: New York, 2005.
[http://dx.doi.org/10.1017/CBO9780511734892]

[33] J.M. Epstein, "Modeling civil violence: an agent-based computational approach", *Proc. Natl. Acad. Sci. USA,* vol. 99, suppl. Suppl. 3, pp. 7243-7250, 2002.
[http://dx.doi.org/10.1073/pnas.092080199] [PMID: 11997450]

[34] R.M. Axelrod, *The complexity of cooperation: Agent-based models of competition and collaboration.* Princeton University Press, 1997.

[35] X. Pan, C.S. Han, and K. Dauber, "A multi-agent based framework for the simulation of human and social behaviors during emergency evacuations", *AI Soc.,* vol. 22, pp. 113-132, 2007. [http://dx.doi.org/10.1007/s00146-007-0126-1]

[36] M.M. Waldrop, *Complexity: The emerging science at the edge of order and chaos.* Simon and Schuster, 1993.

[37] R. Stacey, "Management and the science of complexity: If organizational life is nonlinear, can business strategies prevail?", *Res. Technol. Manag.,* vol. 39, p. 8, 1996. [http://dx.doi.org/10.1080/08956308.1996.11671056]

[38] C. Horn, *A complex systems perspective on communities and tourism : A comparison of two case studies in Kaikoura and Rotorua. Lincoln University.* New Zelandhttps: Canterbury, 2002. hdl.handle.net/10182/1606

[39] M. Batty, "Planning support systems and the new logic of computation", *Reg. Dev. Dialogue,* vol. 16, pp. 1-17, 1995.

[40] F. Wu, and C.J. Webster, "Simulation of land development through the integration of cellular automata and multicriteria evaluation", *Environ. Plann. B,* vol. 25, pp. 103-126, 1998. [http://dx.doi.org/10.1068/b250103]

[41] T.C. Schelling, "Models of Segregation", *Am. Econ. Rev.,* vol. 59, pp. 488-493, 1969.

[42] J.M. Sakoda, "The checkerboard model of social interaction", *J. Math. Sociol.,* vol. 1, pp. 119-132, 1971. [http://dx.doi.org/10.1080/0022250X.1971.9989791]

[43] J.M. Epstein, and R. Axtell, *Growing Artificial Societies: Social Science from the Bottom Up.*Cambridge, MA, 1996.

[44] L. von Bertalanffy, *General system theory: Foundations, development, applications.* Braziller: New York, 1968.

[45] F. Wu, "An experiment on the generic polycentricity of urban growth in a cellular automatic city", *Environ. Plann. B Plann. Des.,* vol. 25, pp. 731-752, 1998. [http://dx.doi.org/10.1068/b250731]

[46] PA Johnson, *Visioning Local Futures: Agent-Based Modeling as a Tourism Planning Support System.* 2009.

[47] J.M. Epstein, "Agent-based computational models and generative social science", *Gener Soc Sci Stud Agent-Based Comput Model,* vol. 4, pp. 4-46, 1999.

[48] A. Crooks, The Use of Agent-Based Modeling for Studying the Social and Physical Environment of Cities, *Complexity and Planning: Systems, Assemblages and Simulations.* Ashgate: Burlington, 2008, pp. 360-393.

[49] S. Franklin, and A. Graesser, Is it an Agent, or just a Program?: A Taxonomy for Autonomous Agents, *Intelligent agents III agent theories, architectures, and languages.* Springer, 1997, pp. 21-35. [http://dx.doi.org/10.1007/BFb0013570]

[50] C.M. Macal, and M.J. North, "Tutorial on agent-based modelling and simulation", *J. Simul.,* vol. 4, pp. 151-162, 2010. [http://dx.doi.org/10.1057/jos.2010.3]

[51] M. Wooldridge, Intelligent agents: The key concepts.*Multi-Agent Systems and Applications II.* Springer, 2002, pp. 3-43. [http://dx.doi.org/10.1007/3-540-45982-0_1]

[52] V. Grimm, U. Berger, and D.L. Deangelis, "The ODD protocol : A review and first update", *Ecol.*

Modell., vol. 221, pp. 2760-2768, 2010.
[http://dx.doi.org/10.1016/j.ecolmodel.2010.08.019]

[53]   G.N. Gilbert, *Agent-based models. Quantitati.* Sage Publications: Los Angeles, 2008.
       http://us.sagepub.com/en-us/nam/agent-based-models/book230292#.WarqLZH-RXU.mendeley
       [http://dx.doi.org/10.4135/9781412983259]

[54]   S.M. Manson, S. Sun, and D. Bonsal, Agent-based modeling and complexity.A.J. Heppenstall, A.T.
       Crooks, L.M. See, and M. Batty, *Agent-based models of geographical systems* Springer, 2012, pp.
       125-139.
       [http://dx.doi.org/10.1007/978-90-481-8927-4_7]

[55]   P-O. Siebers, and U. Aickelin, Introduction to Multi-Agent Simulation.*Encyclopaedia of Decision
       Making and decision support technologies*, 2007, pp. 554-564.

[56]   F. Stonedahl, and U. Wilensky, *NetLogo Virus on a Network modelhttp,* 2008. ccl.northwestern.
       edu/netlogo/models/VirusonaNetwork

[57]   M. Bratman, "Two faces of intention", *Philos. Rev.,* vol. 3, pp. 214-236, 1984.

[58]   M. Bratman, "Plans and resource-bounded practical reasoning", *Comput. Intell.,* vol. 4, pp. 349-355,
       1988.
       [http://dx.doi.org/10.1111/j.1467-8640.1988.tb00284.x]

[59]   B. Head, W. Rand, and U. Wilensky, *NetLogo Traffic Basic Adaptive Individuals,* 2015.
       modelhttp://ccl.northwestern.edu/netlogo/models/TrafficBasicAdaptiveIndividuals

[60]   Y. Zhou, *Path finding model using the A-star algorithm in Netlogo,* 2016. https://github.com/-
       YangZhouCSS/roads

[61]   S. Milgram, "The Small World Problem", *Psychol. Today,* vol. 2, pp. 60-67, 1967.

[62]   D.C. Parker, S.M. Manson, and M.A. Janssen, "Multi-Agent Systems for the Simulation of Land-Use
       and Land-Cover Change: A Review", *Ann. Assoc. Am. Geogr.,* vol. 93, pp. 314-337, 2003.
       [http://dx.doi.org/10.1111/1467-8306.9302004]

[63]   H.A. Simon, "Bounded rationality in social science: Today and tomorrow", *Mind Soc.,* vol. 1, pp. 25-
       39, 2000.
       [http://dx.doi.org/10.1007/BF02512227]

[64]   H.A. Simon, "Theories of bounded rationality", *Decis Organ,* vol. 1, pp. 161-176, 1972.

[65]   H.A. Simon, *Models of Bounded Rationality: Empirically Grounded Economic Reason.* vol. Vol. 3.
       MIT Press: Massachusetts, 1997.

[66]   D.J. Watts, and S.H. Strogatz, "Collective dynamics of 'small-world' networks", *Nature,* vol. 393, no.
       6684, pp. 440-442, 1998.
       [http://dx.doi.org/10.1038/30918] [PMID: 9623998]

[67]   U. Wilensky, *NetLogo Follower model.* http://ccl.northwestern.edu/netlogo/models/Follower (1998).

[68]   I. Boavida-Portugal, C.C. Ferreira, and J. Rocha, "Where to vacation? An agent-based approach to
       modelling tourist decision-making process", *Curr. Issues Tour.,* pp. 1-18, 2015.

[69]   S. Bandini, S. Manzoni, and G. Vizzari, "Agent Based Modeling and Simulation : An Informatics
       Perspective", *J. Artif. Soc. Soc. Simul.,* vol. 12, p. 4, 2009.

[70]   H.R. Gimblett, *Integrating geographic information systems and agent-based technologies for
       modeling and simulating social and ecological phenomena,* Oxford University Press Oxford, UK:
       New York, USA, 2002.

[71]   M. Batty, Approaches to Modelling in GIS: Spatial Representation and Temporal Dynamics.*GIS,
       Spatial Analysis and Modelling.,* D.J. Maguire, M. Batty, M.F. Goodchild, Eds., ESRI Press:
       Redlands, CA, 2005, pp. 41-61.

[72]   M. Batty, "The size, scale, and shape of cities", *Science,* vol. 319, pp. 769-771, 2008.

[73]   H. Couclelis, "' Where has the future gone?' Rethinking the role of integrated land-use models in spatial planning", *Environ. Plann. A,* vol. 37, pp. 1353-1371, 2005.
[http://dx.doi.org/10.1068/a3785]

[74]   A. Crooks, C. Castle, and M. Batty, "Key challenges in agent-based modelling for geo-spatial simulation", *Comput. Environ. Urban Syst.,* vol. 32, pp. 417-430, 2008.
[http://dx.doi.org/10.1016/j.compenvurbsys.2008.09.004]

[75]   P.A. Johnson, and R. Sieber, Agent-Based Modelling : A Dynamic Scenario Planning Approach to Tourism PSS.*Planning Support Systems Best Practice and New Methods.,* S. Geertman, J. Stillwell, Eds., Springer: Netherlands, 2009, pp. 211-226.
[http://dx.doi.org/10.1007/978-1-4020-8952-7_11]

[76]   J.R. McDonald, "Complexity science: an alternative world view for understanding sustainable tourism development", *J. Sustain. Tour.,* vol. 17, pp. 455-471, 2009.
[http://dx.doi.org/10.1080/09669580802495709]

[77]   N. Scott, C. Cooper, and R. Baggio, "Destination Networks", *Ann. Tour. Res.,* vol. 35, pp. 169-188, 2008.
[http://dx.doi.org/10.1016/j.annals.2007.07.004]

[78]   A. Zahra, and C. Ryan, "From chaos to cohesion—Complexity in tourism structures: An analysis of New Zealand's regional tourism organizations", *Tour. Manage.,* vol. 28, pp. 854-862, 2007.
[http://dx.doi.org/10.1016/j.tourman.2006.06.004]

[79]   A.J. Heppenstall, A.T. Crooks, and L.M. See, *Agent-Based Models of Geographical Systems* Dordrecht: Springer: Netherlands, 2012. Epub ahead of print
[http://dx.doi.org/10.1007/978-90-481-8927-4]

[80]   R. Baggio, "Complex tourism systems: a visibility graph approach", *Kybernetes,* vol. 43, pp. 445-461, 2014.
[http://dx.doi.org/10.1108/K-12-2013-0266]

[81]   R. Baggio, "Studying complex tourism systems : a novel approach based on networks derived from a time series", *XIV April Int Acad Conf Econ Soc Dev*

[82]   J. Baggio, Agent-based Modeling and Simulations, *Quantitative Methods in Tourism: A Handbook.,* R. Baggio, J. Klobas, Eds., Channel View Publications: Bristol, 2011, pp. 199-219.

[83]   R. Itami, R. Raulings, and G. MacLaren, "RBSim 2: simulating the complex interactions between human movement and the outdoor recreation environment", *J. Nat. Conserv.,* vol. 11, pp. 278-286, 2003.
[http://dx.doi.org/10.1078/1617-1381-00059]

[84]   P.A. Johnson, and R.E. Sieber, "Negotiating constraints to the adoption of agent-based modeling in tourism planning", *Environ Planning-Part B,* vol. 38, pp. 307-321, 2011.
[http://dx.doi.org/10.1068/b36109]

[85]   P.A. Johnson, and R.E. Sieber, "An individual-based approach to modeling tourism dynamics", *Tour. Anal.,* vol. 15, pp. 517-530, 2010.
[http://dx.doi.org/10.3727/108354210X12889831783198]

[86]   D. Chao, K. Furuta, and T. Kanno, Agent-Based simulation system for supporting sustainable tourism planning. *New Frontiers in Artificial Intelligence.* Springer, 2011, pp. 243-252.
[http://dx.doi.org/10.1007/978-3-642-25655-4_23]

[87]   C. Doscher, K. Moore, and C. Smallman, "An Agent-Based Model of Tourist Movements in New Zealand : Implications for Spatial Yield", *19th International Congress on Modelling and Simulation,* 2011, pp. 12-16. Perth, Australia

[88]   F. Pizzitutti, C.F. Mena, and S.J. Walsh, "Modelling Tourism in the Galapagos Islands", *J. Artif. Soc.*

Soc. Simul., vol. 17, pp. 1-15, 2014.
[http://dx.doi.org/10.18564/jasss.2389]

[89] M. Pons-Pons, P. Johnson, and M. Rosas-Casals, "Modeling climate change effects on winter ski tourism in Andorra", *Clim. Res.,* vol. 54, pp. 197-207, 2012.
[http://dx.doi.org/10.3354/cr01117]

[90] M. Pons, P.A. Johnson, and M. Rosas, "A georeferenced agent-based model to analyze the climate change impacts on ski tourism at a regional scale", *Int. J. Geogr. Inf. Sci.,* vol. 28, pp. 2474-2494, 2014.
[http://dx.doi.org/10.1080/13658816.2014.933481]

[91] S. Balbi, C. Giupponi, and P. Perez, "A spatial agent-based model for assessing strategies of adaptation to climate and tourism demand changes in an alpine tourism destination", *Environ. Model. Softw.,* vol. 45, pp. 29-51, 2013.
[http://dx.doi.org/10.1016/j.envsoft.2012.10.004]

[92] S. Nicholls, B. Amelung, and J. Student, "Agent-Based Modeling: A Powerful Tool for Tourism Researchers", *J. Travel Res.,* 2016. Epub ahead of print
[http://dx.doi.org/10.1177/0047287515620490]

[93] P.A. Longley, M.F. Goodchild, and D.J. Maguire, *Geographic Information Science and Systems* Hoboken: Wiley (John Wiley & Sons, Inc.), 2015.

[94] M.F. Goodchild, GIS and modeling overview.*GIS, spatial analysis, and modeling.,* D.J. Maguire, M. Batty, M.F. Goodchild, Eds., ESRI Press: Redlands, CA, 2005, pp. 1-18.

[95] C. Yu, and D.J. Peuquet, *A GeoAgent-based framework for knowledge-oriented representation: Embracing social rules in GIS,* 2009. Epub ahead of print
[http://dx.doi.org/10.1080/13658810701602104]

[96] J.D. Westervelt, "Computational Approach to Integrating GIS and Agent-Based Modeling", *Integrating Geographic Information Systems and Agent-Based Modeling Techniques,* 2002.

[97] D.J. Maguire, *Implementing spatial analysis and GIS applications for business and service planning.* GIS Bus Serv Plan, 1995, pp. 171-191.

[98] L. Bernard, and T. Krüger, "Integration of GIS and spatio-temporal simulation models: interoperable components for different simulation strategies", *Trans. GIS,* vol. 4, pp. 197-215, 2000.
[http://dx.doi.org/10.1111/1467-9671.00049]

[99] D.J. Maguire, M.F. Goodchild, M. Batty, Ed., *GIS, Spatial Analysis, and Modeling.* Esri Press: Redlands, 2005.

[100] A. Patel, A. Crooks, and N. Koizumi, "Slumulation : An Agent-Based Modeling Approach to Slum Formations", *J. Artif. Soc. Soc. Simul.,* vol. 15, pp. 1-15, 2012.
[http://dx.doi.org/10.18564/jasss.2045]

[101] N. Shoval, B. McKercher, and E. Ng, "Hotel location and tourist activity in cities", *Ann. Tour. Res.,* vol. 38, pp. 1594-1612, 2011.
[http://dx.doi.org/10.1016/j.annals.2011.02.007]

[102] J.C. Hallo, J.A. Beeco, and C. Goetcheus, "GPS as a method for assessing spatial and temporal use distributions of nature-based tourists", *J. Travel Res.,* vol. 51, pp. 591-606, 2012.
[http://dx.doi.org/10.1177/0047287511431325]

[103] F. Girardin, J. Blat, and F. Calabrese, "Digital footprinting: Uncovering tourists with user-generated content", *IEEE Pervasive Comput.,* vol. 7, pp. 36-44, 2008.
[http://dx.doi.org/10.1109/MPRV.2008.71]

[104] D.Z. Sui, "The wikification of GIS and its consequences: Or Angelina Jolie's new tattoo and the future of GIS", *Comput. Environ. Urban Syst.,* vol. 32, pp. 1-5, 2008.
[http://dx.doi.org/10.1016/j.compenvurbsys.2007.12.001]

[105]   N. Shoval, B. McKercher, and A. Birenboim, "The application of a sequence alignment method to the creation of typologies of tourist activity in time and space", *Environ. Plann. B Plann. Des.,* vol. 42, pp. 76-94, 2015.
[http://dx.doi.org/10.1068/b38065]

[106]   P.A. Johnson, R.E. Sieber, and N. Magnien, "Automated web harvesting to collect and analyse user-generated content for tourism", *Curr. Issues Tour.,* vol. 15, pp. 293-299, 2012.
[http://dx.doi.org/10.1080/13683500.2011.555528]

[107]   R.C. Mill, and A.M. Morrison, *The tourism system: An introductory text.* Prentice-Hall, Inc.: Englewood Cliffs, 1985.

[108]   G. Dann, "Anomie, ego-enhancement and tourism", *Ann. Tour. Res.,* vol. 4, pp. 184-194, 1977.
[http://dx.doi.org/10.1016/0160-7383(77)90037-8]

[109]   C. Goossens, "Tourism information and pleasure motivation", *Ann. Tour. Res.,* vol. 27, pp. 301-321, 2000.
[http://dx.doi.org/10.1016/S0160-7383(99)00067-5]

[110]   Y. Yoon, and M. Uysal, "An examination of the effects of motivation and satisfaction on destination loyalty: a structural model", *Tour. Manage.,* vol. 26, pp. 45-56, 2005.
[http://dx.doi.org/10.1016/j.tourman.2003.08.016]

[111]   H. Bansal, and H.A. Eiselt, "Exploratory research of tourist motivations and planning", *Tour. Manage.,* vol. 25, pp. 387-396, 2004.
[http://dx.doi.org/10.1016/S0261-5177(03)00135-3]

[112]   E. Sirakaya, and A.G. Woodside, "Building and testing theories of decision making by travellers", *Tour. Manage.,* vol. 26, pp. 815-832, 2005.
[http://dx.doi.org/10.1016/j.tourman.2004.05.004]

[113]   J. Wong, and C. Yeh, "Tourtist hesitation", *Ann. Tour. Res.,* vol. 36, pp. 6-23, 2009.
[http://dx.doi.org/10.1016/j.annals.2008.09.005]

[114]   A. Decrop, Tourists' decision-making and behavior processes.*Consumer behavior in travel and tourism.,* A. Pizam, Y. Mansfeld, Eds., The Haworth Press, Inc.: New York, USA, 1999, pp. 103-133.

[115]   R.G. Sargent, Simulation model validation.*Simulation and model-based methodologies: an integrative view.* Springer, 1984, pp. 537-555.
[http://dx.doi.org/10.1007/978-3-642-82144-8_19]

[116]   R.G. Sargent, "Verifying and validating simulation models", *Proceedings of the Winter Simulation Conference,* 2014, pp. 118-131

[117]   E.J. Rykiel, "Testing ecological models: the meaning of validation", *Ecol. Modell.,* vol. 90, pp. 229-244, 1996.
[http://dx.doi.org/10.1016/0304-3800(95)00152-2]

[118]   N. Gilbert, and K.G. Troitzsch, *Simulation for the Social Scientist. 2nd editio.* Open University Press: New York, USA, 2005.

[119]   R.C. Mill, and A.M. Morrison, *The tourism system.* Kendall Hunt: Dubuque, IA, 2009.

[120]   S.A. Cohen, G. Prayag, and M. Moital, "Consumer behaviour in tourism: Concepts, influences and opportunities", *Curr. Issues Tour.,* pp. 1-38, 2013.

[121]   D. Kornhauser, U. Wilensky, and W. Rand, "Design Guidelines for Agent Based Model Visualization", *J Artif Soc Soc Simul.* vol. 12.

[122]   B. Mandelbrot, *The Fractal Geometry of Nature. Updated and Augmented.* W. H. Freeman and Company: New York, 1982.

[123]   G. Langran, *Time in Geographic Information Systems. Technical Issues in GIS.* Taylor & Francis: London, 1992.

CHAPTER 3

Spatial Geostatistical Analysis Applied To The Barroso-Alvão Rare-Elements Pegmatite Field (Northern Portugal)

David Silva[1], Alexandre Lima[2,*], Eric Gloaguen[3], Charles Gumiaux[4], Fernando Noronha[2] and **Sarah Deveaud[3]**

[1] *DGAOT, University of Porto, R. Campo Alegre 687, Portugal*

[2] *Institute of Earth Science (ICT), Faculty of Sciences, University of Porto, Porto, Portugal*

[3] *BRGM, ISTO, UMR 7327, av. Claude Guillemin, 45060 Orléans, France*

[4] *ISTO, UMR 7327, Université d'Orléans,- CNRS/INSU - BRGM, Orléans, France*

Abstract: The geological science has been in recent years an excellent playground for GIS applied studies, especially regarding the mineral deposits prospectivity. Other fields of study in the geological science (*e.g.* soil risk management, mining exploitation, geothermal resources…) also took advantages of this geocomputing methodology to extract spatial information. The geoscientist community fairly agrees that interrelations between mineral deposits and certain geological features are observed in the terrain, presenting also a non-random spatial regional distribution pattern in a vast majority of cases. This is where the spatial analysis using geocomputational techniques, in this particular case for rare-elements pegmatites, can be used as a great analytical tool to produce a mapping of mineral potential, or unveil the regional zonation patterns for this type of mineralization. In this study, statistical spatial analyses were performed for the pegmatites to highlight any possible relationship, or lack of it, between them and the surrounding granitic plutons, shear zones or schistose foliations. To accomplish our proposed objectives, the geocomputational method of Distance to Nearest Neighbours (DNN), Ripley's L'-function and pegmatites orientations families were employed to study the spatial distribution pattern of the pegmatites, whereas Euclidean distance and Kernel density distributions aimed the spatial association between these same pegmatites to the various geological features within the study area. The obtained results show: i) Pegmatites spatial distribution following a clustering pattern, presenting the Li-enriched pegmatites a higher rate and extent compared to the total pegmatites, as well as a spatial association with moderate to high pegmatites density; ii) Three distinct families of pegmatites orientation; iii) No statistically significant spatial relationship for the total pegmatites or Li-enriched relatively to the granitic pluton; iv) A regime of deformation within the study area, suggesting the presence of corridors of deformation

* **Corresponding author Alexandre Lima**: Institute of Earth Science (ICT), Department of Geosciences, Environment and Spatial Planning (DGAOT), Faculty of Sciences of Porto University, Portugal; Tel: +351 220402489, Fax: +351 220402490; E-mail: allima@fc.up.pt

Ana Cláudia Teodoro (Ed.)

with NW to NNW orientations; and v) Pegmatites spatial emplacement suggesting shear-zones control.

Keywords: DNN, GIS, Geostatistics, Interpolated foliation, Kernel density, Pegmatites, Ripley's L'-function, Shear-Zones, Spatial analysis, Variogram.

INTRODUCTION

Granitic pegmatites rich in rare elements are currently recognized for their concentration and diversity in ore minerals, both in metallic and industrial minerals [1 - 2]. The importance of pegmatites in our actual society, regarding to the provision of certain types of elements (*e.g.* tantalum (Ta), niobium (Nb) and lithium (Li)), was recognized by the European Union (EU) through the identification of these elements as "Strategic resources" or "Critical raw materials", in part because of the risk of provision or their economic and/or military importance for some European countries [3]. The rare-elements class pegmatites, particularly the LCT (Li-Cs-Ta) subclass [4], are considered the most important economically and best pegmatite class studied to date, presenting the elements Li, Ta, Nb, Be, Sn and Rare-Earth Elements (REE) as the most common critical/strategic elements found in this type of pegmatites [5]. The importance of these rare elements in industrial processes is not very well known by the public, but it is of substantial importance in electronic, automobile or aerospatial industries [6]. Within the LCT-type pegmatites, the most studied bodies regarding their petrogenesis and composition are the giant pegmatites of Tanco, Canada [7] and Greenbushes, Australia [8], however some pegmatites field (*e.g.* Black Hills, USA [9], Ambazac, France [10], Barroso-Alvão, Portugal [11, 12] and Fregeneda-Almendra, Portugal-Spain [13]), also gained scientific interest regarding their petrogenesis, chemical fractionation and spatial emplacement within their pegmatites fields. Up to date, the methodologies used to study the pegmatites where more linked with mineralogical, geochemical and petrological aspects of specific giant pegmatites bodies (*e.g.* Tanco and Greenbushes pegmatites). These results, especially from Tanco, but also from other pegmatites around the world, were applied by London [14] to develop the most used theoretical genetic model of pegmatite associated with magmatic intrusion. This model is widely used in the exploration of industrial minerals or rare-metals bearing pegmatites. However, not all the rare-elements pegmatites field demonstrates evident spatial association or genetic linkage with magmatic intrusions bodies, generally of granitic source (*e.g* [10, 15, 16]). In this type of pegmatites field, the spatial emplacement of the rare-elements pegmatites bodies appears to be controlled by regional shear-zones or faults (*e.g* [10, 16]). To better understand the process controlling the installation and emplacement of LCT pegmatites in non-obvious granite related pegmatite field, a study aiming the spatial relationship between pegmatites bodies and surrounding lithological structures is necessary. This study is based on the premise that pegmatites in the Barroso-Alvão region were not extensively studied, regarding their field scale emplacement or the relationship between pegmatites and surrounding geological structures (fragile/ductile tectonics and granitic lithologies). The previous studies for this area were more focused in a mineralogical and petrographical description of the pegmatites, sometimes containing explanations about the Li mineralized pegmatites localization and emplacement within the Barroso-Alvão pegmatitic field. The spatial relationship methodology is used to better understand some processes regarding the pegmatite's emplacement model, differentiation trends and conception of a genetic model. To achieve these objectives, spatial statistical analysis using geocomputational techniques were applied, especially the Geographic Information System (GIS) technology, to the pegmatites in general and, in more focus, to the mineralized ones. The results are posteriorly interpreted and discussed, to elaborate the most accurate genetic model possible for the regional emplacement of Barroso-Alvão (Portugal) pegmatites.

GEOLOGICAL SETTINGS

The Barroso-Alvão aplitopegmatitic field (Fig. **1**) is situated in the northwestern portion of the Iberian Peninsula, at northern Portugal, in the Trás-os-Montes region. The pegmatitic field area covers partially four sheets of 1:50 000 geological maps of Portugal, which are 6A-Montalegre, 6B-Chaves, 6C-Cabeceiras de Basto, and 6D-Vila Pouca de Aguiar. The studied pegmatite field is located on the occidental portion of the Hesperian massif, a part of the so-called Ibero-Armorican arc. The Iberian Peninsula is affected by the Variscan orogeny starting in the Devonian and continuing into the Carboniferous up to early Permian time. Its structural organization results from three main deformation phases that acted during the Variscan orogeny as proposed by Ribeiro A [17]. The 1st phase (D1), compressive, originates mostly folds with NW-SE axis with vertical axial plan in the Central Iberian Zone (CIZ). The 2nd phase (D2), tangential, induces subhorizontal displacement, and the 3rd phase (D3) covers all the zones developing folds with subvertical axial planes. The pegmatite field is emplaced inside the Galicia Trás-os-Montes Zone (GTMZ), an essentially parautochthonous geotectonic zone using the classification proposed by Julivert [18] for the Iberian Massif, but very close at south to the boundary (the basal thrust) between GTMZ and CIZ. In the oriental sector of the study area is localized the regional fault Régua-Verin (late-D3) presenting an NNE-SSW trend [19].

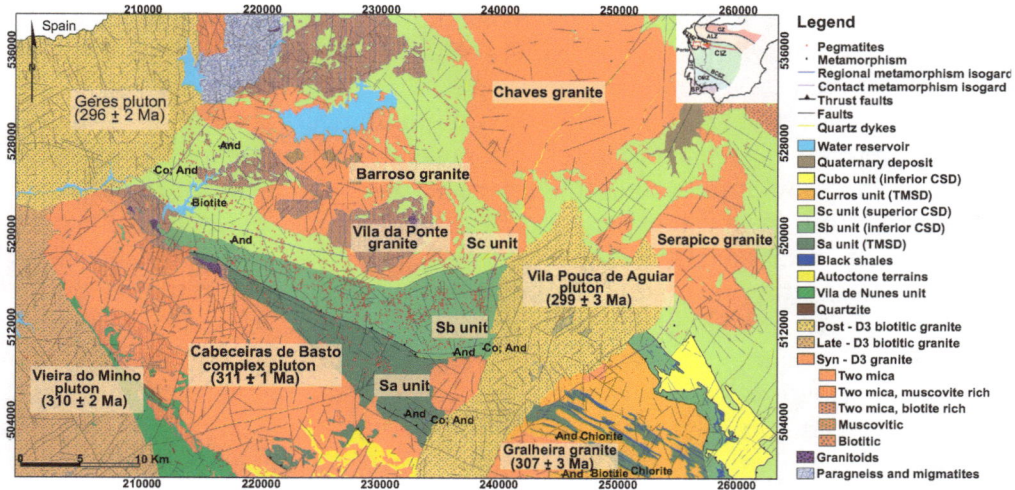

Fig. (1). Simplified geology of the Barroso-Alvão region based on 1:50 000 scale regional geological maps. UTM coordinates according to Lisbon Hayford-Gauss IGeoE datum. Displayed emplacement ages for the various granitic rocks from [20 - 24].

The pegmatites belonging to the Barroso-Alvão field intrude mainly parautochthonous terrains. This one is made of early-Paleozoic metasedimentary and volcano- sedimentary metapelitic, mica schist, and rarely carbonaceous or graphitic schist and is classified in Sa, Sb and Sc unit using lithological and structural criteria [25]. Three superimposed schistosities (S1, S2 and S3) are observed in the metasedimentary rocks consequence of also three deformational phases (D1, D2, D3) originated during the Variscan orogeny that characterizes this zone ([26 - 28]).

The septentrional region of Portugal presents a large diversity of granitoid rocks intruding mainly thick autochthonous and parautochthonous metasedimentary rocks. Taking into account the relationship between granitic rocks and the Variscan orogeny, we can divide the granitoids belonging to the Barroso-Alvão aplitopegmatitic field as syn- to late-D3 granitoids (Syn-D3) defining the pegmatitic field to the north and southwest and post-tectonic granitoids (Post-D3) defining the east and in small amount the northwest limit. The more relevant Syn-D3 granitoids can be individualized in Barroso and Cabeceiras de Basto complex plutons, as for the Post-D3 granitoids the Vila Pouca de Aguiar (VPA) and Gerês plutons are considered the most important ([19, 29, 30]).

The Barroso-Alvão Rare Element Pegmatite Field

This pegmatite field is recognized for its large number of outcropping veins (~2 000) ([11, 12, 31 - 36]) from barren, poorly mineralized to geochemically evolved, and sometimes with low grades of Sn mineralization. The aforementioned recent work of Martins [12] allowed the pegmatite division inside the pegmatitic field in five different groups (Table 1), following the classification of Černý and Ercit [4] as a result of field observations, pegmatite relationships with the host rocks, pegmatites internal structures and mineral geochemistry.

Table 1. Pegmatite type classification and respective mineralogy inside the Barroso-Alvão pegmatitic field.

Pegmatite type	Mineralogy
Intragranitic	Mainly quartz, feldspar, muscovite, biotite and minor tourmaline, beryl and garnet
Barren	The more primitive group; quartz, feldspar, muscovite and minor biotite, apatite, beryl, tourmaline, chlorite, zircon, pyrite and (Ce)-monazite)
Spodumene	Spodumene, Nb-Ta minerals, (F)-apatite, montebrasite, phosphoferrite, chlorite, tourmaline, uraninite and sphalerite
Petalite	Petalite, cassiterite, Nb-Ta minerals, (F)-apatite, montebrasite, pyrite, uraninite, sphalerite and (Ce)-monazite
Lepidolite	Albite, lepidolite, cassiterite and phosphate minerals with Sr-A

METHODOLOGY

The geoscientist community fairly agrees that some mineral deposits present a non- random spatial pattern, mainly because the geneses of those mineral deposits are controlled by the combination of some known geological processes. Between these mineral deposits, some of them show an evident spatial association with certain geologic characteristics, but definitively not with all of them [37]. To tackle these problems, several authors developed or employed GIS methods/techniques to produce prospectivity maps using knowledge-driven and/or data-driven analysis. To achieve the predictive maps, the most usual applied methods include the Boolean methods, weight-of-evidence scores estimation and fuzzy logic approach [38], or algebraic *processus* (*e.g* [39]). Modelling of exploration targets using hybrid methods (fuzzy weight-of-evidence, data-driven fuzzy and neuro-fuzzy modelling) have been proposed to optimize the utilization of both knowledge-driven and data-driven conceptual methods [40, 41]. In this study, the objective was not to create a model of exploring targets *per se* like the cases cited above, but instead, to highlight any potential spatial link between pegmatites and the surrounding structures, especially the ductile shear zones, obliquity with regional schistosity, fault families and granitic units. Therefore, to accomplish our objectives, certain methodologies following some of the work of Deveaud *et al.* [10] where GIS geocomputation were employed. An important validation method used in Deveaud *et al.* [10] and also used in this study, was the use of a statistical method that permits to highlight statistically relevant frequency distributions in the comparative histograms obtained. To do so, the frequency distribution of pegmatites is always compared to a set of points evenly distributed within the same area as the corresponding pegmatites. These evenly distributed points are called reference points and are calculated relatively to the same geological neighboring unit used for the pegmatites. The comparison between pegmatite frequency distribution and the reference frequency distribution permits to categorize the pegmatites distribution into three categories, depending on the pegmatite/reference points ratio. If the pegmatite/reference points frequency distribution ratio is < 1.2, the pegmatite distribution is called a Normal Distribution (*ND*); a ratio value within the range [1.2; 1.5] is a Low Abnormal Distribution (*LAD*), and for a ratio value > 1.5 the pegmatite distribution is called a High Abnormal Distribution (*HAD*).

Spatial Analysis of Pegmatite Bodies

The analysis of spatial point pattern can be achieved using different methodologies. The methods used only the pegmatites as target are: i) Distance to Nearest Neighbor (DNN) [42 - 44]. This statistical method allows obtaining a measure of the spacing of individuals in a population of known density, and the spatial relationship in this population. A ratio (R) is given, representing a measure of the degree to which the distribution pattern of the observed population deviates from random expectation, with a range from 0 (clustered) to 2.1491 (scattered with hexagonal pattern); ii) Ripley's *K*-function [45]. As the DNN, the Ripley's *K*-function, and its derivatives, are suited for analyzing completely mapped spatial point process data and highlights deviations from spatial random distribution of objects at many different distance scales; iii) Euclidean distribution density of pegmatites was used with the intention to highlight the possible relationship between the two groups of pegmatites considered in this study, the totality of pegmatites bodies in the study area and the known Li-enriched pegmatites; iv) Spatial emplacement of pegmatites by orientation families was used, since the orientation of the pegmatites does not present the same orientation in all the study area. To better understand the pegmatites emplacement from the Barroso-Alvão region, it is essential to classify these pegmatites into orientation families, to extract paths or regions of preferential pegmatites emplacement.

Spatial Analysis of Pegmatite Bodies in Relation to Surrounding Granitic Rocks

In this method, the pegmatites are not used as the only target for the geospatial analysis. Here two different geological structures were used: i) a selected set of pegmatite bodies, and ii) the surrounding major syn-D3 granitic bodies in the Barroso-Alvão region (Fig. **2**). The method consists in assigning a Euclidean distance spacing for each individual pegmatite body from the two major granitic bodies in the study area: i) the Barroso pluton and ii) Cabeceiras de Basto granite. This method permits to evaluate if the pegmatites are spatially related, or not, to the surrounding granites.

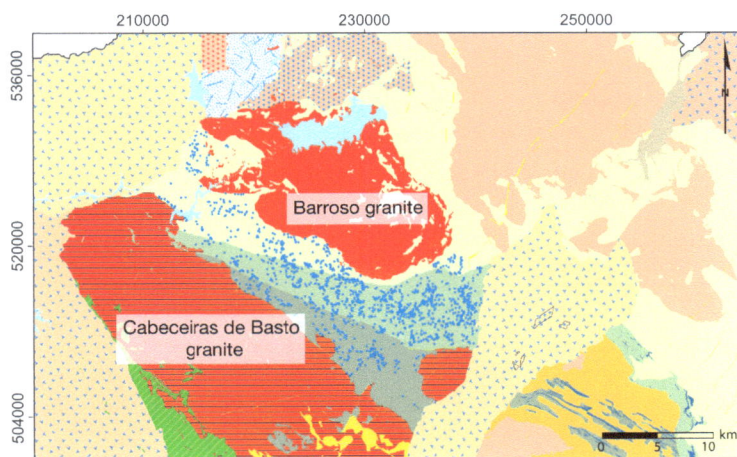

Fig. (2). Map highlighting the two granite from the study area used for the pegmatite distance relationship analysis. The blue points represent all the pegmatite belonging to the constrained study area, later explained in this paper.

Spatial Analysis Relationship of Pegmatite Bodies and Interpolated Metasedimentary Cleavage Trajectories

Following the geostatistical data interpolation developed by Matheron [46], the automatic interpolation methodology developed by Gumiaux *et al.* [47] applies a treatment of directional data to analyze their spatial variations within a selected domain or sector. This methodology uses the *Kriging* method developed by Matheron [46]. The *Kriging* interpolation raster created was used to produce a map of regional foliation orientation trajectories and to calculate the angular difference between the local interpolated cleavage striking and the pegmatites orientation.

Dataset

The Barroso-Alvão area was chosen as object of this study, in part, because of factors already described in the previous works (*e.g.* large population of pegmatites, presence of Li minerals, similar age and tectonic setting than Monts d'Ambazac pegmatite field in Massif Central, France [10]), but also because of the large GIS-applicable data available and published for this region (*e.g.* mineralogy, geochemistry, orientation and localization of pegmatite dykes) (Table 2). This amalgam of data was published over years in form of maps, papers and thesis by several authors (*e.g* ([11, 12, 29, 31, 32, 34 - 36]).

Table 2. Information used for the compilation of the Barroso-Alvão aplitopegmatite field GIS.

Theme	Content
Geological maps of Portugal	1:50.000 scale: 6A-Montalegre, 6B-Chaves, 6C-Cabeceiras de Basto, and 6D-Vila Pouca de Aguia; 1:200.000 scale: sheet nº 2
Geomaps	Granitic pegmatites: the state of the art – field trip guidebook [48]; Prospeg project – pegmatite remote sensing and mapping. Final report [49]; Structural map in Ramos [50].
DEM	Landsat 7 ETM+ cover (Band 1, 2, 4)
Satellite Imagery	World Imagery MapServer from Esri ArcGIS
Mineral Inventory	Geological maps of the studied area and respective reports; Personal data acquired *in situ* at Barroso-Alvão aplitopegmatitic field; Dra. Tânia Martins (Porto University) doctoral program; Prof. Alexandre Lima (Porto University) personal database.

Another aspect that is relevant to the choice of Barroso-Alvão field is the past artisanal and actual modern industrial exploitation in pegmatites focused in this region. To compute the different spatial statistical analyses, layers of information containing the localization and characteristics of the Barroso-Alvão region geological structures were compiled. All the layers and their content were georeferenced using the Lisbon Hayford-Gauss IGeoE datum (ESRI: 102164) for precise results. These vector layers are: i) Point layer (Pegmatites); ii) Line layer (Faults and thrust faults; Metasediment foliation), and iii) Polygon layer (Lithology).

Geological Layer

The lithological layer was created with the aim to unify the different geology, using their description and symbols, from the four geological maps at 1:50 000 scale. This task was not easily made, because of the almost forty years that separate the youngest and the oldest maps. To difficult this task even more, the 6B-Chaves map (oldest) and the 6D-Vila Pouca de Aguiar map (youngest) are edge-to-edge with each other. The criteria for rock classification changed over the forty years that separate these two geological maps. To solve this problem, the geological map of Portugal at 1:200 000 scale was used to simplify this issue between sectors in the 6B and 6D map and also between 6B and 6A map. The classification for the different lithologies was done using the classification proposed in the geological maps 6A, 6C and 6D. The granitic rocks were classified using a structural feature (Sin-, Late or Post-D3) combined with a mineralogical classification representing the mica content. The metasediments were classified following their structural level, composition and deformation. The continuity of the lithological classification over the years between the maps 6A, 6C and 6D is assured, because of the participation of one of the authors (Prof. Fernando Noronha) has the mapping manager for these three geological maps.

Pegmatite Layer

For the geological maps 1:50 000, the data was compiled digitalizing all the pegmatites visible in the maps (Fig. **3a**). The pegmatites in the four maps are represented as lines, allowing extracting the orientation of the structure, or as pegmatites masses, representing pegmatites with sub-horizontal dipping. The difficulty was to digitize, as lines, these pegmatites masses to obtain their orientations. Later, the segments digitalized were transformed into points based on the centroid of each segment. The same process was used for the maps extracted from the publications cited above, except for the map in Granitic pegmatites: the state of the art – field trip guidebook [48] where the pegmatites were directly represented as points. In this exception, the method applied was to digitalize the localization of these pegmatites directly as a layer point, without the extraction of orientation.

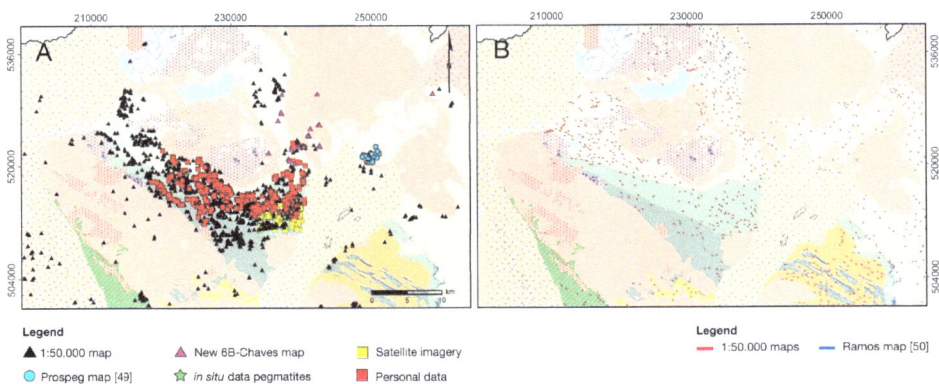

Fig. (3). (**A**) Point layer representing the complete set of pegmatite data, classified by compilation source; (**B**) Line layer representing the complete set of foliation data, classified by compilation source.

Satellite imagery was used to complete a specific area of the Barroso-Alvão pegmatitic field. This area, near, at west, to Vila Pouca de Aguiar pluton, is poorly represented with pegmatites in the maps described anteriorly. This fact is due to the step topology and dense vegetation, which makes nearly impossible to execute a proper mapping campaign. Although these difficulties one of the authors (Prof. Alexandre Lima) as known this area as containing mineralized pegmatites, reason why the aforementioned area was targeted using satellite data for pegmatites digitalization. The digitalization process was the same as anteriorly described. The only difference was that the old artisanal exploitations were also targeted for digitalization.

Concerning the personal databases, the database from Prof. Alexandre Lima was a compilation of fieldwork from Alexandre Lima himself in Barroso-Alvão region and in 6B-Chaves geological map region for the upgrade of this map, and from Dra. Tânia Martins (Porto University) during her doctoral program. The points from the personal database were acquired *in situ,* using a GPS device, during a field trip.

Metasediment Foliation Layer

The digitalization process was similar to the previous pegmatites and faults digitalization. Each foliation observed in the maps was digitize as a line, and is dip and dip direction characteristics compiled into the database layer (Fig. **3b**). The orientation was later extracted using the extension EasyCalculate in the ArcGIS (ESRI) software. Not all the foliations were digitize, only the S3 foliation, because the S1 and S2 dip are frequently lower than 45° and the S3 foliation are always sub-vertical. This choice was also made because of the chronological order of the regional cleavage, in which S3 is the last foliation created within Barroso-Alvão region, and the dense and highly visible presence of these foliations in the hinge zone of regional folds. Finally, cleavage trajectory maps only apply for a single deformation event and different generations of cleavage must be treated separately. The use of the map from Ramos [50] for foliation digitalization helped to validate our assumption about the S3 foliation from 6-A and 6-B maps.

DATA PREPARATION

Barroso-Alvão Region and Pegmatite Analysis

The Barroso-Alvão region shows a total of 1924 pegmatites occurrence, which 1828 (95%) of them are emplaced in the metasediment host lithology and 96 (5%) in the remainder lithologies. From these residual 96 pegmatites are emplaced in other lithologies, 93 (4.8%) are situated in granitic lithology and 3 are located (0.2%) in migmatites. The ratio of pegmatites emplaced in granite *versus* metasediment is approximatively 1:20. Regarding the Li-mineralized pegmatites, in the 1924 total pegmatites, 78 (4%) of them are mineralized with Li minerals as spodumene, petalite and lepidolite as the most representative phases. These mineral phases, regarding the major mineral phase present, were used to classify the pegmatites into three classes: i) Spodumene pegmatites (54 occurrences), ii) Petalite pegmatites (22 occurrences) and iii) Lepidolite pegmatites (2 occurrences). All the remainder pegmatites are unknown respecting their mineralization content.

The study area as a dimension of 2 560 km^2, in which the N-S extent is 40 km and the E-W extent is 64 km. Inside the study area, the granitoid lithology emerges as the dominant lithology with 1 600.25 km^2 (62.5%), followed by the metasedimentary lithology with 854.35 km^2 (33.4%). Because of the disparity observed in for the lithological localization of pegmatites, the spatial distribution analysis of pegmatites was carried using only the pegmatites within the metasediment host lithology (1828 occurrences). After classifying the pegmatites in metasediment as the only pegmatites bodies for spatial distribution analysis, the distribution of the pegmatite dataset continued too spatially dispersed. For this reason, a second constriction was applied to the pegmatites within metasediments lithologies in this case to a specific area within the Barroso-Alvão region. This reduction in the study area was necessary in order to obtain more precise results in the pegmatites distribution, particularly because of a visually obvious large concentration of pegmatite (1627 occurrences (89%)) in this specific sector of Barroso-Alvão pegmatitic field. Moreover, the fact that all the 78 Li-mineralized pegmatites were emplaced in this sector also strengthened the choice made.

Schistosity Cleavage Interpolation Analysis

As described previously, the schistosity cleavage interpolation followed the geostatistical data interpolation developed by Matheron [46] and the automatic interpolation methodology developed by Gumiaux *et al.* (2003). The Gumiaux *et al.* [47] methodology applies a treatment of orientation data to analyze their spatial variations within a selected domain or sector. The *Kriging* method allows interpolating spatially distributed data point, following the principle that spatially close points tend to exhibit more similar values than distant points. Due to their variation from 0° to 360°, classic statistical calculation used in non-circular algebraic data, like the *Kriging* method, is impossible to apply here. To bypass this problem, instead of using algebraic difference between the orientation points, Gumiaux *et al.* [47] obtain the *Kriging* interpolation using direction cosines, where each orientation of the data points is represented by its two direction cosines. This direction cosine method was used in the first part of the *Kriging* interpolation that consists in quantifying the spatial variation of the data set by the calculation of experimental variograms.

From the original set of 793 foliation orientation digitalized, only the digitalized cleavages within the study area (269 measures) where used. From these, only 14 (5%) of them display gently dipping (below 45°) and trajectories in map view can thus be used to highlight spatial variations in the strain field. From this, only the foliation with a minimum inclination of 45° were used for the interpolation process. The 14 shallow dipping foliation measurements are evenly distributed over the study area, not biasing the calculation of the foliation trajectories interpolation.

For the foliation within the study area, the orientation distribution follows a bell shaped Gaussian curve. This main normal distribution allows calculating the variogram of the study area (Fig. **4**).

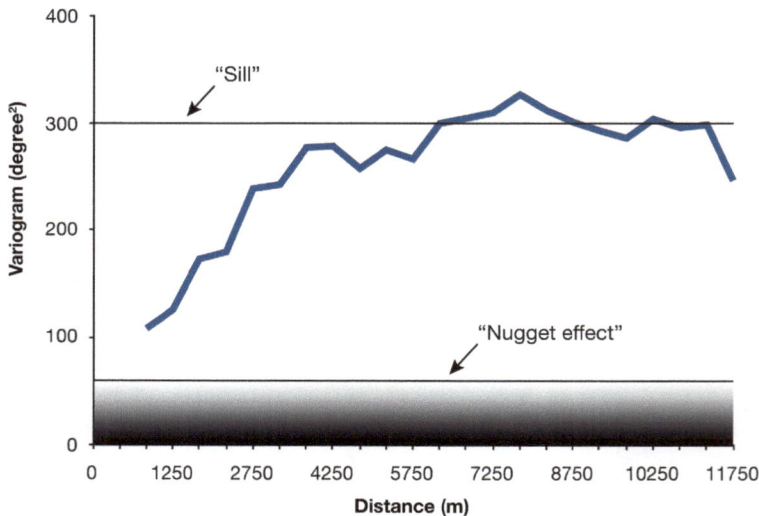

Fig. (4). Experimental variogram calculated using the Barroso-Alvão metasediment foliations orientation.

The experimental variogram was calculated using 255 orientation data points, presenting a maximum distance calculation of 12 km, which corresponds approximately at half diagonal distance of the study area, and a lag distance interval of 250 m. The spatial correlation between data pairs is observed up to approx. 5 500 m (range value), the "sill" (corresponding to the averaged maximum and constant values reached with increasing distance) at approx. 300 degree^2 and the "nugget effect" (background noise data or effect at scales inferior than the study area) at 60 degree^2, representing 20% of the total variance. The experimental trend of the variogram follows an exponential theoretical model.

Using the range value from the experimental variogram as the search radius for the *Kriging*, the interpolation of orientation data was later performed in a GIS software.

RESULTS

Spatial Distribution Analysis of Pegmatites

Distance to Nearest Neighbor (DNN)

Analysis of the frequency distribution of the computed DNN for the 1 627 pegmatites occurrences, shows an average neighbor distance of 143 m, a minimum distance of 5.6 m and a maximum distance of 1 613 m. To obtain the calculation area necessary to complete the DNN methodology, a buffer was created using all the 1 627 pegmatites (Fig. **5a-c**). This buffer was calculated using the same maximum distance (1 613 m), obtained in the frequency distribution of all pegmatites described above, to limit the buffer maximum distance calculation. The same approach was also carried out for all the Li-enriched pegmatites (Fig. **5a** and **d**). In these pegmatites, the average neighbor distance obtained was 287 m, with a minimum distance of 44 m and a maximum distance of 2 891 m.

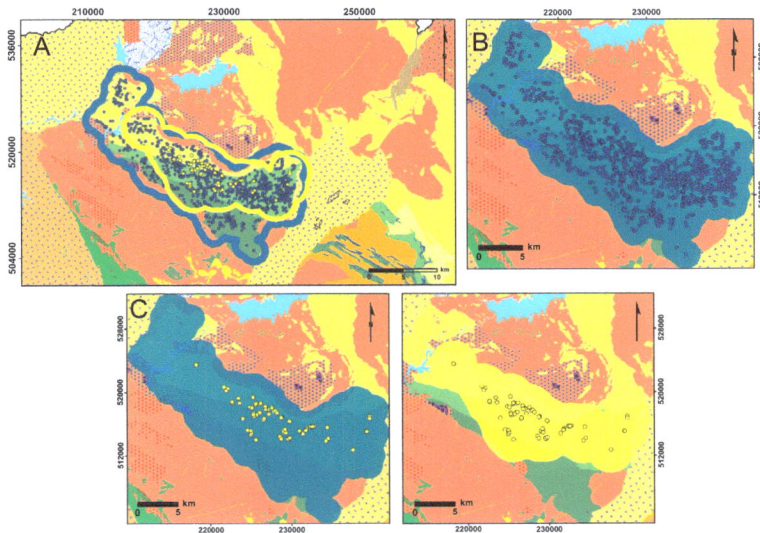

Fig. (5). The two buffers used for the DNN R ratio and Ripley's *K*-function calculation; (**A**) Localization of the buffers and groups of pegmatites in the study area; (**B**) All pegmatites group and respective buffer created from it; (**C**) Spatial pattern of Li-enriched pegmatites inside the buffer created from all pegmatites group; (**D**) Li-enriched pegmatites group and respective buffer. Blue buffer - All pegmatites group; Yellow buffer – Li-enriched pegmatites group.

In Fig. (**6a** and **b**), it is possible to observe the close neighborhood relationship existing between pegmatites. In these histograms, the large majority of pegmatites are located in the left side of the graphs, where the class distances of pegmatites nearest neighbor are smaller (143 m and 287 m for all pegmatites and Li-enriched pegmatites respectively). In the two groups of pegmatites analyzed, all pegmatites and Li-enriched pegmatites, the calculated DNN median values, 100 m and 168 m respectively, were smaller than the average DNN values for each group. These results show that more than 50% of the pegmatites occurrences, in his respective group, are at smaller distance to is neighbor than the average DNN calculated.

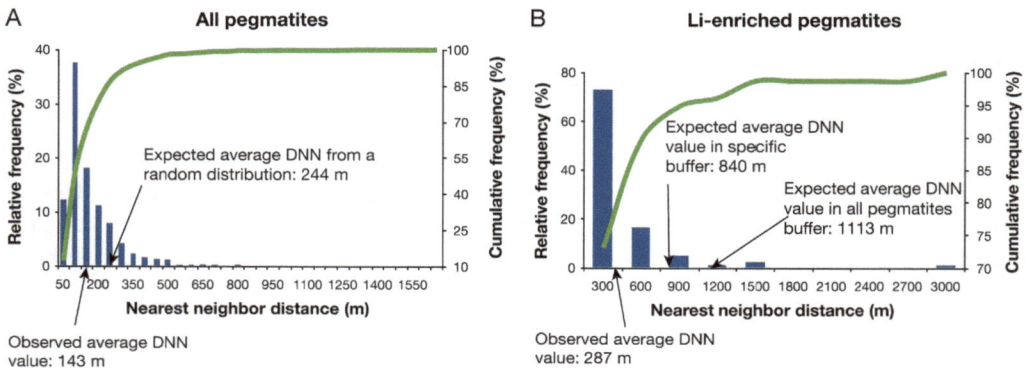

Fig. (6). Relative and cumulative frequency distribution of the DNN for the pegmatites considered in the Barroso-Alvão aplitopegmatitic field; (**A**) All pegmatites group DNN calculation histogram; (**B**) Li- enriched group DNN calculation histogram.

The calculated ratio (R) obtained for the total 1 627 pegmatites in the high-density sector, using the respective buffer of 386.35 km^2 (Fig. **5b**) as calculation area, is 0.59. Using the same area as previously, an identical calculation was performed for the Li-enriched pegmatites occurrences (Fig. **5c**). This methodology was applied to determine if the Li-enriched pegmatites followed the same, or a different path of distribution, as the total 1 627 pegmatites. The obtained ratio (R) of 0.26 for the Li-enriched pegmatites demonstrates a probable separated path of clustering relatively to the total pegmatite for the same area of calculation. A second R ratio calculation was applied to the Li-enriched pegmatites to strengthen the evidence of higher degree of clustering for these pegmatites. In this calculation, the difference to the previous ratio (R) calculation was in the buffer used as area of calculation (220.31 km^2; Fig. **5d**). The buffer was created directly from the Li-enriched pegmatites using the maximum distance of 2 891 m between these pegmatites, to limit the buffer maximum distance calculation. The ratio (R) of 0.34 obtained, even evidencing a smaller degree of clustering than the previous 0.26 ratio (R) by reducing the area of the reference zone, permits to confirm the segregated and higher degree of clustering previously evidenced for Li-enriched pegmatites relatively to the total pegmatites in the high pegmatite density sector.

Ripley's K-function

The Ripley's *K*-function was applied only for the pegmatites within the high-density zone (see Fig. **5a** and **b**), following the same conclusion for the choice made previously described in the DNN calculation. The *K*-function statistic is very sensitive to the size of the study area. Identical arrangements of points can exhibit clustering or scattering depending on the size of the study area enclosing them. Therefore, to avoid these distribution discrepancies, the buffers created for all the pegmatites and for the Li-enriched pegmatites anteriorly used to calculate the DNN ratio (R), were also used here in the *K*-function (Fig. **5b** and **d**).

Two different L'-function were computed, one in each buffer for their respective pegmatite. Regarding the L'-function computed for all pegmatites (Fig. **7**), the L'(r) value increases (*i.e.* clustering) with distance and scatters immediately after r= 3 600 m and becomes random at 7 200 m. During the increase of L'(r), two distinctive clustering rates are observed: i) a moderate clustering rate up to 2 000 m, ii) a low clustering rate from 2 000 m up to 3 600 m.

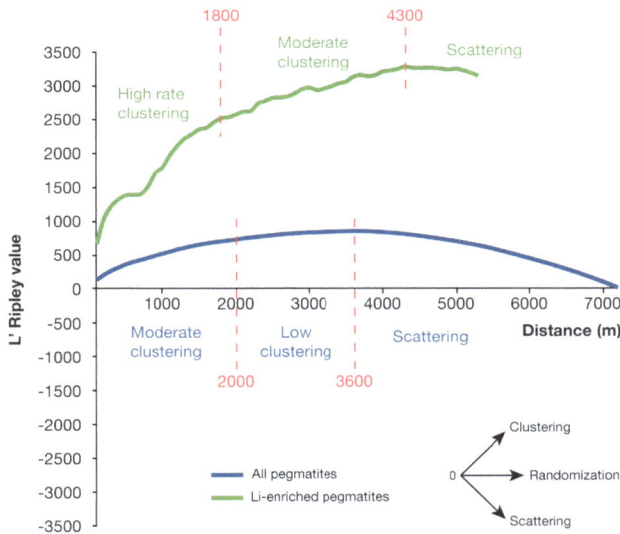

Fig. (7). Ripley's L'-function computed to detect spatial deviation from a homogeneous Poisson distribution. All pegmatites present a Ripley L' function value = 730 for the moderate clustering rate at a distance of 2000 m and a L' function value = 854 for the low clustering rate at a distance of 3600 m. Li mineralized pegmatites present a Ripley L' function value = 2516 for the high clustering rate at a distance of 1800 m and a L' function value = 3288 for the moderate clustering rate at a distance of 3600 m.

The L'-function computed for the Li-enriched pegmatites presents different paths and clustering rates, compared with the computed values for all pegmatites. The first difference is in the maximum extent of spatial point pattern calculation. For the Li-enriched pegmatites, the maximum extent is 5 300 m compared with the higher 7 200 m for all pegmatites. This smaller extent for Li-enriched pegmatites shows a smaller dispersion in the terrain for these pegmatites in comparison with all pegmatites. The second difference is the almost total clustering path observed and the rates of clustering presented by the Li-enriched pegmatites. The L'(r) value increases up to 4 300 m, scattering from there until 5 200 m (*i.e.* the maximum calculation distance extent for these pegmatites). During the increase of L'(r), two distinctive rates of clustering are observed: i) high rate of clustering up to 1 800 m, ii) moderate clustering rate from 1 800 m up to 4 300 m. The results obtained from Ripley's K-function permitted to validate the results from DNN calculation, showing that the Li-enriched presents a distinctive path and rate of clustering compared to the remainder pegmatites in the same zone.

Euclidean Distribution Density of Pegmatites

The density of all pegmatites was computed in ArcGIS 10.2 (ESRI) software using the *kernel density calculation* function. This function calculates the magnitude per unit area, in our case km^2, from the all pegmatites points using a kernel function to fit a smoothly tapered surface to each pegmatite. Later, the spatial localization of the Li-enriched group was compared with the all pegmatites density calculation anteriorly described (Fig. **8**).

The analysis of the histogram representing the Li-enriched pegmatites group distribution shows that comparatively to the all pegmatites group kernel density (Fig. **9a**), the Li-enriched pegmatites frequency distribution extends from the all pegmatites density per km^2 value 2 to 24, whereas the reference distribution extends from 0 to 26. The first class in the histogram, representing the lowest pegmatites density, does not contain any Li-enriched pegmatite, but for the reference distribution it is in this class where the highest frequency value is observed (>30%). The first two classes where Li-enriched pegmatites are present, [2 - 4] and [4 - 6], shows that the reference distribution values are much higher than the values for the pegmatites distribution, classifying therefore these two classes for pegmatite distribution as *ND*. In contrast, beyond all pegmatites density of 6, the Li-enriched pegmatites distribution is always sufficiently higher to be classified as *LAD* for the classes [6 - 8] and [8 - 10], and as *HAD* for the classes from [10 - 12] to [22 - 24]. The only exception is the last class [24 - 26] where no Li-enriched pegmatites are present. These results highlight that the Li-enriched pegmatite group are apparently spatially linked to higher pegmatites density values, and especially with density values within the range 10 to 24.

However, none of the Li-enriched pegmatites are emplaced within the maximum class density [24 - 26].

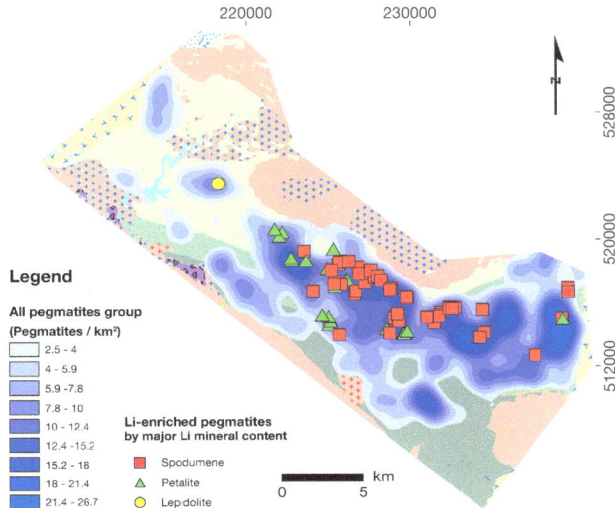

Fig. (8). Map of the kernel density (km) for the all pegmatites group, and spatial distribution of the Li-enriched pegmatites group and respective major Li mineral content (spodumene, petalite and lepidolite) pegmatites sub-groups.

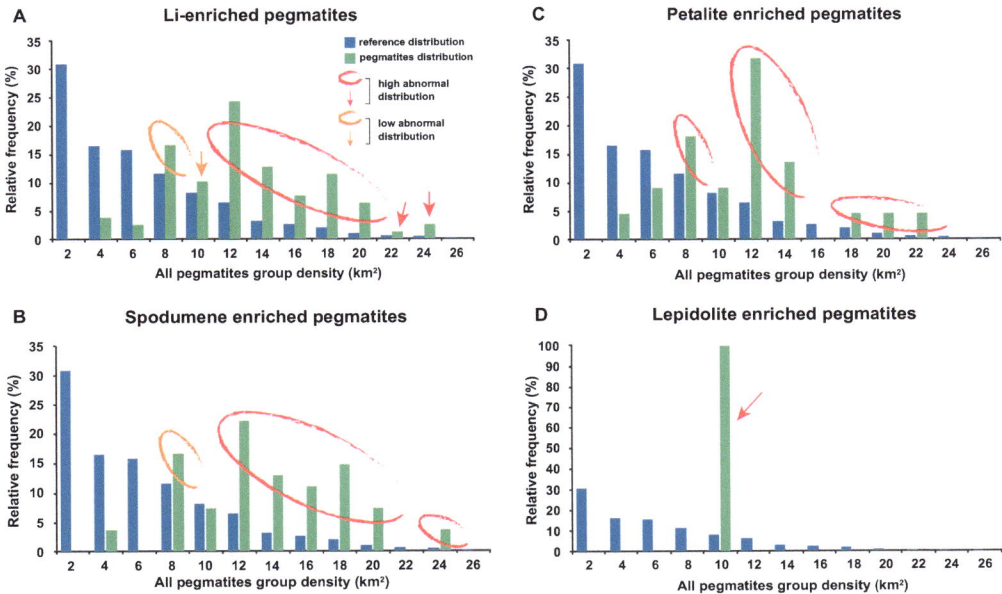

Fig. (9). Histograms representing the Li-enriched pegmatites group and respective sub-groups distribution comparatively to the all pegmatites group kernel density.

The histogram for the spodumene-enriched pegmatites (Fig. **9b**) present in the Li-enriched pegmatite group, shows the same frequency distribution extent as the total Li- enriched pegmatites, from pegmatite density 2 to 24. However, the spodumene pegmatites are absent in the class [4 - 6] and [20 - 22] from this range. The first density class where spodumene pegmatites are present shows a frequency distribution strongly lower compared to the reference frequency distribution, classifying this class as *ND*. Identically to the total Li- enriched pegmatites, it is beyond the pegmatite density value of 6 that the spodumene pegmatites distributions become statistically abnormal, exception to the class [8 - 10]. Within the range density 6 to 24, the density class [6 - 8] is the only classified as *LAD*. All the other classes from this range presenting abnormal distributions show spodumene pegmatites with a *HAD* type.

The petalite-enriched pegmatites histogram (Fig. **9c**) shows a distribution extent slightly shorter than the previous total Li-enriched and spodumene pegmatites, from 2 to 22. Other differences observed comparatively to the Li-enriched pegmatite distribution are the absence of petalite pegmatites in the density class [14 - 16], and in absence of *LAD* pegmatites type for this sub-group. It is beyond the pegmatite density of 6, as seen for the total Li- enriched and spodumene pegmatites, that the abnormal pegmatites distributions are observed. Different to the previous histograms, the petalite-enriched pegmatites presents a *HAD* for all the abnormal distribution beyond the pegmatite density of 6, exception for the class [8 - 10] with an *ND* type and the aforementioned absence of petalite pegmatites in the class [14 - 16]. The spodumene and petalite-enriched pegmatites present approximately a similar spatial distribution as observed in the total Li-enriched. Comparing the spodumene and petalite-enriched pegmatites, the petalite pegmatites, as a whole, are apparently emplaced in lower pegmatites density relatively to the spodumene pegmatites. The lepidolite pegmatites sub-group presents only two occurrences in all the study area, with almost the same spatial emplacement. Because of that reduced number of occurrences and almost the same localization for the two occurrences, the lepidolite pegmatites frequency distribution histogram (Fig. **9d**) presents lepidolite pegmatites only in the [8 - 10] class of pegmatite density. However, even presenting a reduced number of occurrences, these pegmatites show apparently the same spatial correlation with the higher pegmatite value density observed in the previous pegmatites group and sub-groups.

Spatial Emplacement of Pegmatites by Striking Families

To better understand the pegmatites emplacement from the Barroso-Alvão region, it is essential to classify these pegmatites into striking families, in order to extract structural trends of preferential pegmatites emplacement. The classification into striking families permits to visualize: i) a range of pegmatites orientation without the background noise from the other families, ii) better understand if the different families of pegmatites follow the same emplacement mechanism, and iii) the interaction between the different families of pegmatites. The pegmatites grouping into strike families were achieved using the 1 616 pegmatites orientation from the all pegmatites group database. In the database, the number of pegmatites orientation is slightly inferior from the total number of pegmatites in the all pegmatites group because of 11 pegmatites, essentially Li-enriched pegmatites, do not present orientation data. These pegmatites were represented in maps and extracted as points without data about their orientation. All the 1 616 values were transformed from 0°-360° to 0°-180° to facilitate the visualization of the histogram for the frequency distribution of pegmatites orientation (Fig. **10**).

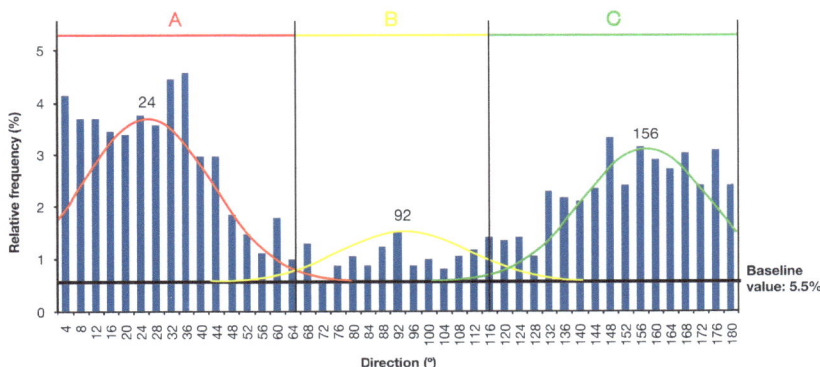

Fig. (10). Histogram representing the domainal distribution of the pegmatites orientation by families.

To obtain the best-fit decomposition calculation of the frequency distribution histogram, the PeakFit software was used. This software computed the deconvolution of the values using a Gaussian type function and a rather low base level value of 0.6% (representing the noise in the pegmatites orientation data set). The distribution of the pegmatites orientation was entirely decomposed into three classes families: i) A class family (n = 775) from N0° to N63° (including N180° as directional data are of circular type) with a maximum at N24°; ii) B class family (n = 216) from N64° to N113° with a maximum at N92°; iii) C class family (n = 625) from N114° to N179° with a maximum fixed at N156°.

After the calculation of the three pegmatites orientation families, these three classes of orientation were applied to the pegmatites layer to create maps displaying only one family at a time (Fig. **11**). In the A class family pegmatites orientation map (Fig. **11a**), a spatial distribution of this pegmatite family covering approximately all the study area were observed, with higher density cluster of A family pegmatite localized in the eastern part of the study area. This sector is also where the highest densities of pegmatites are found in all Barroso-Alvão region. It is also observed that the A family pegmatites show a preference (88% of the pegmatites) for the emplacement in upper stratigraphic levels (Sb and Sc unit from the Carrazedo structural domain). The B class family map (Fig. **11b**) shows a pegmatite spatial distribution to be almost entirely concentrated in the central and eastern sector of the study area, with a preference for the upper stratigraphic levels of the Sb and Sc metasedimentary units (83% of the pegmatites). Finally, in the C class family map (Fig. **11c**) it was observed that these pegmatites cover almost the entirety of study area. Different to the A class family, the higher density distribution for the C class family is observed in the central sector and not in the east sector. Other differences observed are in the respectively higher (22%) and smaller (16%) spatial emplacement of the pegmatites within the Sa and Sc unit compared to the previous families. It is also in this pegmatite family where the pegmatites appear to follow more clearly an alignment, especially in the central sector of the study area.

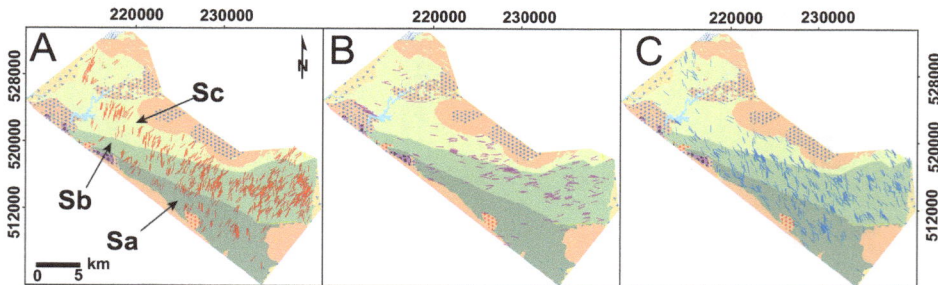

Fig. (11). Maps showing the spatial distribution and orientation of the all pegmatites group divided by family of pegmatites orientation: (**A**) Map of the A class family of pegmatites orientation. (**B**) Map of the B class family of pegmatites orientation. (**C**) Map of the C class family of pegmatites orientation.

The three classes of pegmatites orientation were later applied to the Li-enriched pegmatite group, creating, as previously described for the all pegmatites group, maps displaying individually the pegmatites orientation families A, B and C (Fig. **12**). The map of the Li-enriched pegmatites A class family orientation (Fig. **12a**) shows a narrower spatial extent of emplacement, compared to the distribution from the all pegmatites A family pegmatites map. Here the clusters of higher densities are localized in the east sector of the study area, as for the A Li-enriched pegmatites they are preferentially localized in the central sector. The Li-enriched pegmatites concentration in the central sector of the study area is also observed in the B and C families of Li-enriched pegmatites (Fig. **12b** and **c**), as well as the apparent preference for the upper metasedimentary units, with almost 100% of the Li-enriched pegmatites emplaced in the Sb and Sc units.

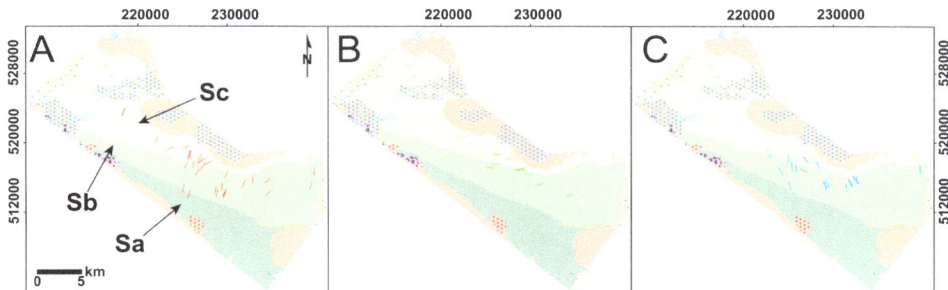

Fig. (12). Maps showing the spatial distribution and orientation of the Li-enriched pegmatites group divided by family of pegmatites orientation: (**A**) Map of the A class family of pegmatites orientation. (**B**) Map of the B class family of pegmatites orientation. (**C**) Map of the C class family of pegmatites orientation.

Spatial Analysis of Pegmatite Bodies in Relation to Surrounding Granitic Rocks

Distance Analysis Between Pegmatites and Barroso Granitic Pluton

The all pegmatites group spatial distribution extends from 0 to 12 000 m, whereas the reference point distribution is slightly larger, from 0 to 14 100 m (Fig. **13a**). The total pegmatites present an average distance from the Barroso pluton of 3 685 m. The first two classes of distance, from [0 to 600 m, show that the frequency distribution for the reference point in this range of distance is always superior than the pegmatites distribution, representing 5% of the total pegmatites. For this range, the frequency distribution between pegmatites and reference points show also an almost perfect anti-correlation. The range of distance from [900-1 200 m] to [4 500-4 800 m] is where the majority of pegmatites are emplaced (62%) and it is also in this range where the pegmatites are observed to display the majority of *LAD* and *HAD* type in all the study area. The pegmatites presenting *LAD* type are present in the classes of [900 to 1 500 m], [2 700 to 3 300 m], [3 900-4 200 m], [5 700-6 000 m] and [6 300-6 600 m], representing 29% of the total pegmatites. The *HAD* type is only present in four class of distance, [2 400-2 700 m] and [3 300 to 3 900 m] and [4 500 to 4 800 m], representing 23% of the total pegmatites.

The spodumene-enriched pegmatites have a spatial extension up to 6 000 m, which is 50% smaller than the extension of the all pegmatites group (Fig. **13b**). The average distance from the Barroso pluton is 1 916 m. *HAD* is present from [600 to 2 700 m], and in the isolated class [3 300-3 600 m]. Further than 3 600 m, the pegmatites present frequency values smaller or insignificantly higher than the reference points. The spodumene-enriched pegmatites show a closer relationship to the Barroso granitic pluton than the all pegmatites group.

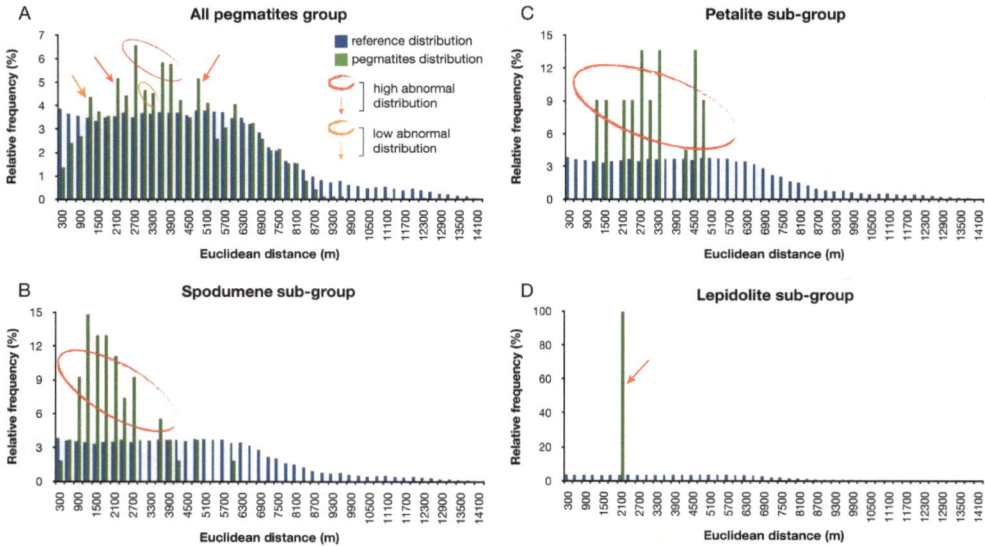

Fig. (13). Histograms representing the all pegmatites group and Li-enriched pegmatites sub-groups distance analysis from the Barroso granitic pluton.

The petalite-enriched pegmatites present an extension from 600 to 4 800 m with an average distance from the Barroso pluton of 2 793 m (Fig. **13c**). The petalite-enriched pegmatites follows clearly a *HAD* type starting from the class [1 200-1 500 m]. This *HAD* path is observed in the range [900-4 500 m] of extension with some exception were the petalite-enriched pegmatites are absent. Only in two class, in the second and the last classes where the pegmatites appear, it is observed *LAD* type. In 35% of the extent, 50% of the total petalite pegmatites sub-group overlaps the spodumene-enriched pegmatites *HAD* classes type.

The lepidolite sub-group pegmatites are observed only in the [1 200 to 1 800 m] distance class, locating this sub-group within the range distance where the two previous pegmatites sub-groups shows their preferential *HAD* paths (Fig. **13d**).

Distance Analysis Between Pegmatites and Cabeceiras De Basto Granitic Pluton

The all pegmatites group spatial distribution extends from 0 to 8 750 m, whereas the reference point distribution is more extensive in almost 3 000 m, from 0 to 11 500 m (Fig. **14a**). The total pegmatites present an average distance from the Cabeceiras de Basto pluton of 3 793 m. From the class of distance [2 250-2 500 m] to [6 000-6 250 m] all the pegmatite frequency values are higher than the reference points, with the exception of the class [2 750-3 000 m] presenting a ratio of 0.92. 68% of the pegmatites are emplaced in the range of distance from the Cabeceiras de Basto granitic pluton described above. It is also in this range where all the pegmatites with *LAD* and *HAD* type are observed. 55% of the total pegmatite presents an anomalous distribution path in the study area. The pegmatites presenting *LAD* type have a more extensive class of spatial extension, [3 000 to 6 250 m], compared to the pegmatites presenting *HAD*, [4 000 to 5 750 m]. From [6 250 m], no class presents pegmatites with frequency values higher than the reference points.

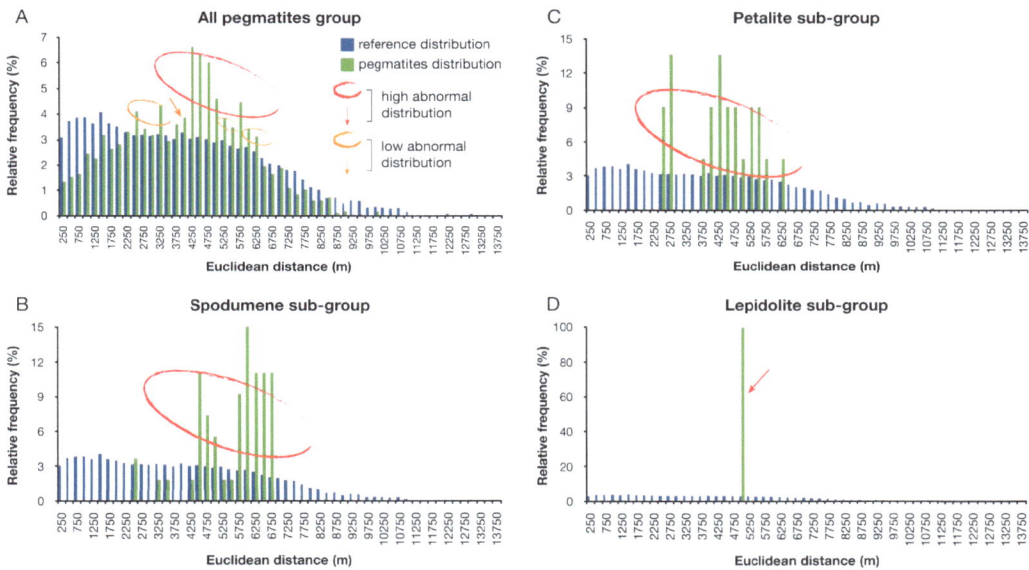

Fig. (14). Histograms representing the all pegmatites group and Li-enriched pegmatites sub-groups distance analysis from the Cabeceiras de Basto granitic pluton.

The spodumene-enriched pegmatites present an extension from 2 000 to 6 750 m with an average distance from the Cabeceiras de Basto pluton of 5 407 m (Fig. **14b**). From the starting class of [2 250-2 500 m] to [3 750-4 000 m], only 7% of this type of pegmatite is represented, despite this range representing 30% of the total spodumene-enriched pegmatites extent. In this range, it is also observed that the reference points values are always higher than the pegmatite frequency values. The remainder 70% of the spodumene-enriched pegmatites are emplaced in the range distance from class [4 000-4 250 m] to the maximum extent 6 750 m following clearly a *HAD* path, exception to the *LAD* class [4 000-4 250 m] and the distance class [5 000 to 5 500 m] presenting pegmatites frequency values under the reference points. The spodumene-enriched pegmatites show an overall distribution more distant from the Cabeceiras de Basto pluton than the all pegmatites group.

The petalite-enriched pegmatites have a spatial extension up to 6 250 m and an average distance from the Cabeceiras de Basto pluton of 4 076 m (Fig. **14c**). As the spodumene-enriched pegmatites, the first pegmatite appearance occurs above the 2 km distance from the granite, in the class [2 250-2 500 m]. This first observed emplacement of petalite-enriched pegmatites marks the first range, from 2 250 to 2 750 m, where is observed petalite pegmatites following a *HAD* path. The second *HAD* path is localized up from 3 500 m to the maximum extent of 6 250 m, with the exception of the class [5 250 to 6 000 m] where no petalite pegmatite is present. 50% of the total *HAD* observed in the petalite-enriched pegmatites are emplaced closer to the pluton than the *HAD* range presented by the spodumene-enriched pegmatites.

The lepidolite sub-group pegmatites are observed only in the [4 500-4 750 m] class of distance, locating this sub-group within the range distance where the two previous pegmatites sub-groups shows preferential *HAD* paths (Fig. **14d**).

Obliquity of Pegmatite Bodies and Interpolated Metasedimentary Cleavage Schistosity

One of the principal objectives of the foliation orientation interpolation was to create a cleavage trajectory map (Fig. **15a**) that allows the visualization and better understanding of the regional deformation occurred in the study area. Trajectories following an approximately constant direction in all the study area were observed. This observation is strengthened by the histogram of the directional foliation data (Fig. **15b**), showing an almost normal striking distribution with a maximum N109° direction. However, the direction of the trajectories in the extremities of the study area, close to the post-tectonic granites pluton of Gerês and VPA, changes relatively to the regional overall schistosity. Although following an approximately overall homogeneous direction, the schistosity trajectories present small changes in their directions creating an apparent undulation or sigmoidal path in all the study area.

Using the interpolated raster of schistosity for the Barroso-Alvão aplitopegmatitic field, it was possible to generate a histogram showing the relationship between the pegmatites orientation and the calculated schistosity for all the study area Fig. **15c**). This histogram shows the frequency of the absolute strike angular difference between the pegmatite orientation and the interpolated schistosity orientation extracted at the same spatial emplacement of the pegmatite occurrence.

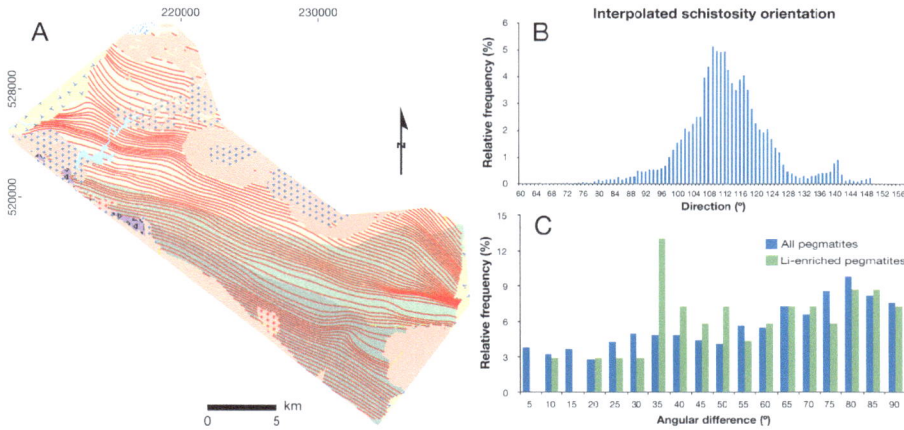

Fig. (15). (**A**) Map showing the interpolated S3 schistosity trajectories within the study area; (**B**) Histogram of the interpolated schistosity orientation for the study area; (**C**) Histogram representing the frequency of the absolute angular difference between the pegmatite orientation and the interpolated schistosity orientation.

The absolute angular difference was applied for the all pegmatites and the Li-enriched pegmatites group. In the all pegmatites group, 48% of the total all pegmatites are emplaced in a range of angular differences with the interpolated foliation orientation from 60° up to the maximum 90°. Admitting a maximum angular difference of 15°, only 11% of the pegmatites are considered as following the metasediment schistosity. The Li-enriched pegmatite group shows a similar frequency distribution for the angular range [60°-90°], with 45% of them to be within this range. Although this resemblance of frequency for this range, below the angular difference of 60°, the Li-enriched pegmatites group presents higher frequency value in the range [35°-50°] (33% of the Li-enriched pegmatites) and only 3% below 15°, whereas the all pegmatites group presents a distribution more evenly distributed for the angular differences bellow 60°.

SYNTHESIS AND INTERPRETATION

In this chapter, the results obtained from the pegmatites spatial statistical analyses relatively to their i) spatial distribution, ii) Euclidean distance relationship to the syn-tectonic plutons of Barroso and Cabeceiras de Basto, and iii) relationship with interpolated foliation orientation, are here interpreted. The main goal was to highlight the most probable mechanism of pegmatite emplacement for the Barroso-Alvão aplitopegmatitic field.

Two major groups of pegmatites were considered in this study. A first group containing all the pegmatites for the study area, and a second containing only the pegmatites knew to be mineralized with the lithium minerals spodumene, petalite or lepidolite, subdivided into sub-groups within the Li-enriched group according to this criterion. The large majority of the pegmatites are emplaced within the metasediments lithology, despite the predominance of granitic lithology in the Barroso-Alvão region.

The statistical analyses concerning the spatial distribution of the pegmatites occurrences, for now considering only the Distance to Neighbor Distance (DNN) and the Ripley's K-function computation, showed that the pegmatites in the study area are emplaced following a clustered distribution. Although this pegmatite clustering, the rate and ratio of clustering for the all pegmatites group and the Li-enriched group is not equal. The two ratios (R) of 0.26 and 0.34 for the Li-enriched pegmatite group, compared to the 0.56 ratio (R) from the all pegmatites group, shows that a segregated and higher degree of clustering is present in the Li-enriched pegmatites group. The Ripley's K-function reinforced the previous DNN result showing that separated rate of clustering is present for the two groups of pegmatites. It is observed up to 1 800 m an especially high rate of clustering for the Li-enriched group, while the all pegmatites group presents a much lower rate of clustering up to 3 500 m followed by scattered distribution. The Li-enriched group also differs in this last characteristic of distribution from the all pegmatites group, showing clustering in all the 4 300 m range distribution.

Within the analysis of the spatial distribution of pegmatites, the spatial density distribution of the all pegmatites was performed with the aim to understand the spatial relationship between this mentioned density and the emplacement of the Li-enriched pegmatite group. In this analysis, the Li-enriched pegmatites, as a group, show a spatial relationship with the higher densities of pegmatites. Individually, the sub-groups of Li-enriched pegmatites demonstrate the same spatial relationship with the higher pegmatites density. These results show an apparent contradiction to the models of fractionated crystallization of pegmatites from parental granite. In these models, the more fractionated pegmatites are generally more distant from the barren pegmatites, and not, as observed for the more evolved pegmatites in Barroso-Alvão, in the higher densities of pegmatites.

The last results obtained for the analysis of the spatial distribution of pegmatites were the families of pegmatites preferential orientation for the two groups of pegmatites. Three families were obtained, family A (N0° to N63°), B (N64° to N113°) and C (N114° to N179°). All the families, either in the all pegmatites group or Li-enriched group, show a preferential emplacement in the upper stratigraphic sub-units' Sb and Sc of the Carrazedo structural domain, as well as a preference to be emplaced in the central and eastern sector of the study area. From all the families, the family C is the one showing the more evident alignments of the pegmatites. Two preferential directions are observed, one ~NNW-SSE more evident in the western part of the central sector, and a second following a ~NW-SE direction in the central and eastern sector. These alignments suggest that corridors of preferential pegmatites emplacement following these two directions are possibly present in the study area. The maximum N24° and N156° orientation values for the orientation pegmatites family A and C respectively, corresponding to a minimum angular difference between them of 48°, as well as the visual observation of these two families in the western part of the central sector, suggest a possible integration of these pegmatites into the same model of emplacement.

In the Euclidean pegmatites distance relationship from the syn-tectonic plutons of Barroso and Cabeceiras de Basto, and taking into account the parental granitic models presented by Černý [51] and London [52] as comparison, the expected higher distribution around the pluton and farthest distance from the sub-types spodumene, petalite and lepidolite are not observed for the Barroso granitic pluton. It is apparently the opposite that occurs, with the all pegmatites group abnormal distributions classes to be more distant from the pluton, and the spodumene, petalite and lepidolite sub-groups at similar or closer distance of the pluton. The only aspect slightly concordant with the above granitic parental models is the average higher petalite sub-group distance from the pluton, relatively to the spodumene sub- group.

Similar to the distance analysis between the pegmatite and the Barroso granitic pluton, it is suggested that the emplacement of the pegmatites relatively to Cabeceiras de Basto pluton distance does not follow the parental granitic model as proposed by Černý [51] and London [52]. Also similar to Barroso pluton, the pegmatites are preferentially emplaced at higher distance from the Cabeceiras de Basto granite. The Li-enriched pegmatites group suggest also an emplacement that contradicts the granite parental model, with: i) a pegmatite emplacement at similar or closer distance from the granite compared to the all pegmatites group, ii) the inverted distance fractionation of the sub-groups and iii) an overlapped distribution, in some part of the extent, of the spodumene, petalite and lepidolite-enriched pegmatites.

Finally, the interpolation of metasedimentary foliation orientation allowed the creation of a map of cleavage trajectories, and the statistical analysis of the strike angular difference between the pegmatites and the interpolated foliation direction. In the map of cleavage trajectories, it is observed an apparent undulation or sigmoidal path in the computed trajectories, suggesting a plane strain deformation with an apparent dextral movement that bend the schistosity within the Barroso-Alvão aplitopegmatitic field. The strike angular difference between the pegmatites orientation and the interpolated foliation histogram shows that the all pegmatites group demonstrate a preference to be emplaced within a range of 60° to 90° differences from the metasediment schistosity. Admitting a maximum angular difference of 15°, only 11% of the pegmatites are considered as following the metasediment schistosity. The Li-enriched pegmatites group shows a similar trend, but also a high preference to be emplaced at 35° to 50° relatively to the schistosity. These results show that the pegmatites are preferentially emplaced at angle to the metasediment schistosity, suggesting that the majority of the pegmatites were emplaced after the D3 deformational phase.

Altogether, these results suggest that the pegmatites spatial emplacement in Barroso-Alvão aplitopegmatitic field followed shear-zones presenting ~NW-SE and ~NNW-SSE orientations. Following this line of thought, we continued to try to find these proposed shear zones using ASTER imagery. The methodology consisted in the interpretation of terrain lineations in the satellite imagery using different image processing (*e.g.* solar exposition, aspect, slope). These lineations where lately included in a map from the study area containing the pegmatites by groups and faults by families, to observe the relationship between them (Fig. **16**).

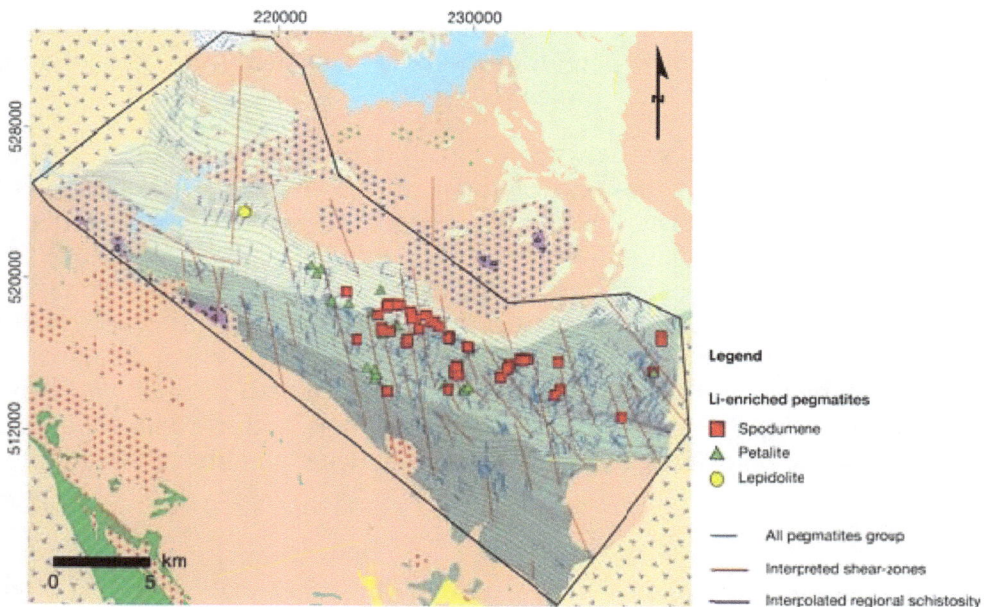

Fig. (16). Map showing the spatial emplacement of all the pegmatites and Li-enriched pegmatites relatively to the interpreted shear-zones and the interpolated regional schistosity.

The interpreted shear zones displayed in Fig. (**16**) show on some occasion a superposed localization to beta family faults, following almost the same direction.

This aspect of the interpreted shear zones is especially well observed in the western part of the central sector of the study area, presenting also in this sector a certain spatial emplacement rhythmicity concerning the distance between them. In this same western sector, it is observed a shear zone that presents a visually obvious spatial relationship with several Li-enriched pegmatites. This spatial relationship is also observed with the pegmatites from the C-family of orientation. The corridors of deformation between the interpreted shear zones present numerous pegmatites following the A-family of pegmatites orientation, suggesting that these pegmatites were preferentially emplaced within these corridors. These shear zones and associated corridors of deformation probably created channel-ways within the metasedimentary units that permitted the emplacement of the pegmatitic melts precursor of the pegmatites bodies. The B-family of pegmatite orientation does not present a particular relationship with the interpreted shear zones, following preferentially orientations similar to the interpolated trajectory of schistosity. Therefore, the emplacement of the B-family pegmatites orientation is probably correlated with the metasedimentary foliations, and possibly, at least, synchronous with the last phase of regional deformation. These visual interpretations showing an association between the pegmatites and the interpreted shear-zones are reinforced with the empirical results from the histogram representing the pegmatites distance under 1000 m from the interpreted shear-zones (Fig. 17). In this histogram, it is possible to observe that the pegmatites presents LAD in distance bellow 350 m for the all pegmatites group, and HAD and LAD in the distance 150 m to 250 m, 400 m and 550 m to 600 m for the Li-enriched pegmatites, showing a close relationship between pegmatites and the interpreted shear-zones.

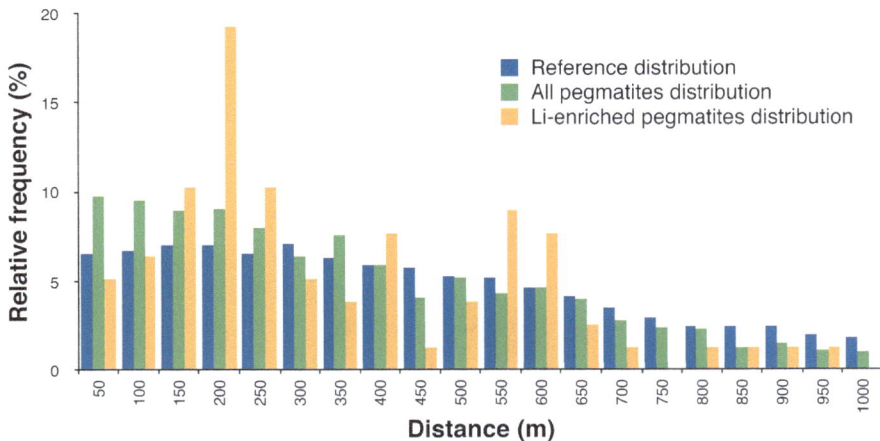

Fig. (17). Histogram showing the distance relationship of the Barroso-Alvão pegmatites under 1000 m from the interpreted shear-zones obtained using ASTER imagery.

DISCUSSION

The methodology applied in this study has brought new evidence concerning i) the difficulty to relate the Barroso-Alvão pegmatites with the Černý [51] and London [52] parental granite model, and ii) a model of pegmatites emplacement related with shear- zones oriented from NW-SE to NNW-SSE.

In the first point, the pegmatites distance from the specialized in Li, Sn and W Cabeceiras de Basto granite, and from the Barroso granite showed an anti-correlation not compatible with the parental granite pegmatites model proposed by Černý and London. The fact that the Li-enriched pegmatites showed preferential emplacement within high density of pegmatites also contradicts the London and Černý model, which predicts that the pegmatites more fractionated, and therefore more chemically evolved, are distant from the higher densities of pegmatites. Previously to our study, Martins *et al.* [35, 36] obtained similar conclusions to our results with their mineralogical chemical evolution for the Barroso-Alvão pegmatites, in which they do not observe a strong pegmatite chemical evolution from the aforementioned granite.

The second point is related to the mechanism of pegmatite emplacement in Barroso-Alvão aplitopegmatitic field and presents some similarities with the model of pegmatites emplacement proposed by Demartis *et al.* [16] for the Comechingones pegmatites field in Argentina. In their model of pegmatites emplacement, the pegmatites are emplaced in spaces developed in a shear-zone according to the Riedel model of fracturation. Our results show that the pegmatites in Barroso-Alvão appear to be emplaced in association with shear corridors. Extrapolating the Riedel fracturation for the Barroso-Alvão pegmatites emplacement, the C-family of pegmatite orientation is proposed to be emplaced and oriented following the C-shear zones or principal plane of deformation. Following the Riedel fracturation, the ~45° between the A and C-family of pegmatite orientation suggest that the A-family of pegmatite were emplaced in tensional fractures perpendicular to the minimum horizontal compression in Barroso-Alvão region, and highly discordant to the metasediment foliation. Due to their orientation, the pegmatites included in the B-family of pegmatite orientation are the only pegmatites that cannot be integrated into this Riedel fracturation model of emplacement. We contend that the emplacement of these pegmatites was controlled by the metasediment foliation. This mechanism of emplacement is not exclusive to the B-family of pegmatite orientation, since some pegmatites from the C-family of pegmatite orientation also present orientation similar to the metasediment direction in some parts of the study area. The highly diverse orientation of the metasediment foliation is a factor which complexify the interpretation of the mechanism of pegmatites emplacement, allowing possible mixture of process of pegmatites emplacement for the B and C-family of pegmatite orientation.

Concerning the Li-enriched pegmatites, statistical computations show that the average distance from the interpreted shear-zones (327 m) is inferior to the distance presented by the all pegmatites group (474 m). These results, integrated within a model of pegmatite emplacement associated with shear-zones corridors, appear to evidence that the Li-enriched pegmatites are younger in comparison to the all pegmatites group. The same results are observed for the spodumene and petalite pegmatites. The spodumene pegmatites present a slightly higher distance (334 m) from the C-shear planes in comparison to the petalite pegmatites (290 m) and an emplacement in a more central position within the corridor of deformation, suggesting that the spodumene pegmatites were emplaced before the petalite pegmatites. These results, although not presenting a strong statistical evidence, are in concordance with the higher fractionation of the Li-enriched and petalite pegmatites, therefore younger, in comparison respectively to the all pegmatites and spodumene pegmatites proposed by Martins *et al.* [35, 36].

A careful interpretation of the results is proposed since all the computational methodologies applied in this study were purely made in two dimensions, not taking into account the pegmatites emplacement in association with the topography of the Barroso-Alvão region.

Compiling all the observations and results obtained in this study, a graphical model of pegmatites emplacement for the Barroso-Alvão aplitopegmatitic field is here proposed (Fig. **18**).

Considering these new results, we contend that the mineralized pegmatites are emplaced preferentially within upper metasedimentary structural levels (Sb and Sc units), closer to the Barroso granite but at distance higher than 600 m from this granite, and spatially close to the interpreted shear-zones oriented from ~NW-SE to ~NNW-SSE acquired using satellite imagery. We propose that these results provide a practical and accurate prospection model for the mineralized pegmatite within the Barroso-Alvão pegmatite field.

CONCLUDING REMARKS

In this study, the spatial statistical analysis using GIS methodologies allowed to follow an approach of studying the pegmatites emplacement that is not easily found in the literature, compared to the more "classical" geological methods like the petrography, geochemistry or geochronology of pegmatites. Different to these last methods, the spatial statistical analysis presents a special focus in determining the spatial emplacement relationship between pegmatites and the surrounding structures (*e.g.* granites, faults, schistosity), while the "classical" methods are more suited to link the pegmatites to the original source and the degree of chemical fractionation. Because of these differences, the GIS methodologies and the more "classical" methods are ideal to be used as a whole, complementing each other deficiencies.

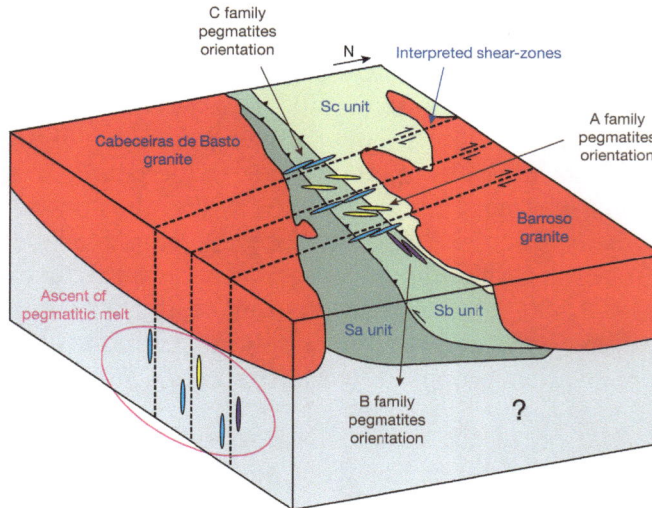

Fig. (18). Simple emplacement model for the pegmatites in the Barroso-Alvão aplitopegmatitic field.

CONSENT FOR PUBLICATION

Not applicable.

CONFLICT OF INTEREST

The author (editor) declares no conflict of interest, financial or otherwise.

ACKNOWLEDGEMENTS

The authors acknowledge the funding provided by the Institute of Earth Sciences (ICT), under contract with FCT (the Portuguese Science and Technology Foundation). The authors also thank the support of the BRGM and the University of Orleans in this manuscript.

REFERENCES

[1]   A.S. Glover, W.Z. Rogers, and J.E. Barton, "Granitic pegmatites: storehouses of industrial minerals", *Elements,* vol. 8, pp. 269-273, 2012.
       [http://dx.doi.org/10.2113/gselements.8.4.269]

[2]   R.L. Linnen, M. Van Lichtervelde, and P. Černý, "Granitic pegmatites as sources of strategic metals",

Elements, vol. 8, pp. 275-280, 2012.
[http://dx.doi.org/10.2113/gselements.8.4.275]

[3]   EU ad hoc working group on defining raw materials, "Report on critical raw materials for the EU", In: *RMSG, DGEI, EC,* 2014.

[4]   P. Černý, and T.S. Ercit, "The classification of granitic pegmatites revisited", *Can. Mineral.,* vol. 43, pp. 2005-2026, 2005.
[http://dx.doi.org/10.2113/gscanmin.43.6.2005]

[5]   D. London, "Granitic pegmatites: an assessment of current concepts and directions for the future", *Lithos,* vol. 80, pp. 281-303, 2005.
[http://dx.doi.org/10.1016/j.lithos.2004.02.009]

[6]   P. Černý, Petrogenesis of granitic pegmatites.*Granitic Pegmatites in Science and Industry. Mineral. Assoc. Can., Short Course Handbook,* P. Černý, Ed., vol. 8. , 1982, pp. 405-461.

[7]   D. London, "Magmatic-hydrothermal transition in the Tanco rare-element pegmatite: evidence from fluid inclusions and phase-equilibrium experiment", *Am. Mineral.,* vol. 71, pp. 376-395, 1986.

[8]   M.I. Hatcher, and B.C. Bolitho, The Greenbushes pegmatite, southwest Western Australia.*Granitic Pegmatites in Science and Industry. Mineral. Assoc. Can., Short Course Handbook,* P. Černý, Ed., vol. 8. , 1982, pp. 513-525.

[9]   J.J. Norton, and J.A. Redden, "Relations of zoned pegmatites to other pegmatites, granite, and metamorphic rocks, in the southern Black Hills, South Dakota", *Am. Mineral.,* vol. 75, pp. 631-655, 1990.

[10]   S. Deveaud, C. Gumiaux, E. Gloaguen, and Y. Branquet, "Spatial statistical analysis applied to rare-element LCT-type pegmatite fields: an original to constrain faults-pegmatites-granites relationships", *J. Geosci. (Prague),* vol. 58, pp. 163-182, 2013.
[http://dx.doi.org/10.3190/jgeosci.141]

[11]   A. Lima, *"Estrutura, mineralogia e génese dos filões aplitopegmatíticos com espodumena na região do Barroso-Alvão (Norte de Portugal)",* PhD dissertation. Portugal: University of Porto, p. 212, 2000.

[12]   T. Martins, *"Multidisciplinary study of pegmatites and associated Li and Sn-Nb-Ta mineralisation from the Barroso-Alvão region",* PhD dissertation. Portugal: University of Porto, p. 196, 2009.

[13]   R. Vieira, *"Rare-elements aplopegmatites from Almendra (V.N. de Foz-Côa) and Barca d'Alva (Figueira Castelo Rodrigo) regions. Aplitopegmatitic field of Fregeneda-Almendra",* PhD dissertation. Portugal: University of Porto, 2010.

[14]   D. London, "Pegmatites. Canadian Mineralogist", *Special Publication,* vol. 10, p. 347, 2008.

[15]   I.H. Henderson, and P.M. Ihlen, "Emplacement of polygeneration pegmatites in relation to Sveco-Norwegian contractional tectonics: examples from southern Norway", *Precambrian Res.,* vol. 133, pp. 207-222, 2004.
[http://dx.doi.org/10.1016/j.precamres.2004.05.011]

[16]   M. Demartis, L.P. Pinotti, and J.E. Coniglio, "Ascent and emplacement of pegmatitic melts in a major reverse shear zone (Sierra de Cordoba, Argentina)", *J. Struct. Geol.,* vol. 33, pp. 1334-1346, 2011.
[http://dx.doi.org/10.1016/j.jsg.2011.06.008]

[17]   A. Ribeiro, Evolução geodinâmica de Portugal: os ciclos ante-mesozóicos.*Geologia de Portugal.,* R. Dias, A. Araújo, P. Terrinha, J.C. Kullberg, Eds., vol. Vol. 1. Escolar Editora: Lisboa, 2013, pp. 15-57.

[18]   M. Julivert, J.M. Fontbote, A. Ribeiro, and L.E. Conde, "Mapa tectónico de la Península Ibérica Y Baleares E1: 1.000.000. Memoria explicativa. Publicación I.G.M", *E,* p. 101, 1974.

[19]   H. Sant'ovaia, J. Bouchez, F. Noronha, D. Leblanc, and J.L. Vigneresse, "Composite-laccolith emplacement of the post-tectonic Vila Pouca de Aguiar granite pluton (northern Portugal): a combined AMS and gravity study", *Trans. R. Soc. Edinb. Earth Sci.,* vol. 91, pp. 123-137, 2000.
[http://dx.doi.org/10.1017/S026359330000732X]

[20] A. Almeida, J. Leterrier, F. Noronha, and J.M. Bertrand, "U-Pb zircon and monazite geochronology of the Hercynian two-mica granite composite pluton of Cabeceiras de Basto (Northern Portugal)", *Comptes Rendus de l'Académie des Sciences,* vol. 326, pp. 779-785, 1998.

[21] G. Dias, J. Leterrier, A. Mendes, P. Simões, and J. Bertrand, "U-PB zircon and monazite geochronology of post-collisional Hercynian granitoids from the Central Iberian Zone (Northern Portugal)", *Lithos,* vol. 45, pp. 349-369, 1998.
[http://dx.doi.org/10.1016/S0024-4937(98)00039-5]

[22] H. Martins, H. Sant'ovaia, and F. Noronha, "Genesis and emplacement of felsic Variscan plutons within a deep crustal lineation, the Penacova-Régua-Verín fault: An integrated geophysics and geochemical study (NW Iberian Peninsula)", *Lithos,* vol. 111, pp. 142-155, 2009.
[http://dx.doi.org/10.1016/j.lithos.2008.10.018]

[23] H. Martins, H. Sant'ovaia, and F. Noronha, "Late-Variscan emplacement and genesis of the Vieira do Minho composite pluton, Central Iberian Zone: Constraints from U–Pb zircon geochronology, AMS data and Sr–Nd–O isotope geochemistry", *Lithos,* vol. 162-163, pp. 221-235, 2013.
[http://dx.doi.org/10.1016/j.lithos.2013.01.001]

[24] H. Priem, and E. den Tex, "Tracing crustal evolution in the NW Iberian Peninsula through the Rb-Sr and U-Pb systematics of Palaeozoic granitoids: a review", *Phys. Earth Planet. Inter.,* vol. 35, pp. 121-130, 1984.
[http://dx.doi.org/10.1016/0031-9201(84)90038-4]

[25] M.A. Ribeiro, F. Noronha, and M. Cuney, *Importância do estudo litogeoquímico na caracterização das unidades tectono-estratigráficas do parautóctone da zona Galiza Média-Trás-os-Montes [CD-ROM].* Ciências da Terra (UNL): Lisboa, 2003, pp. B89-B92.

[26] F. Noronha, J.M. Ramos, J. Rebelo, A. Ribeiro, and M.L. Ribeiro, "Essai de corrélation des phases de déformation hercyniennes dans le NW de la P.I", *Leid. Geol. Meded.,* vol. 52, no. 1, pp. 87-91, 1981.

[27] E. Pereira, A. Ribeiro, and C. Meireles, "Variscan shear zones and control of Sn-W, Au, U mineralizations in the Central-Iberian Zone in Portugal", *Cuaderno Laboratorio Xeolóxico de Laxe,* vol. 18, pp. 89-119, 1993.

[28] R. Dias, and A. Ribeiro, "The Ibero-Armorican Arc: A collision effect against an irregular continent?", *Tectonophysics,* vol. 246, pp. 113-128, 1995.
[http://dx.doi.org/10.1016/0040-1951(94)00253-6]

[29] F. Noronha, and M.L. Ribeiro, *Notícia explicativa da Carta geológica de Portugal na escala 1:50000. Folha 6A - Montalegre.* Instituto Geológico e Mineiro: Lisboa, 1983, p. 30.

[30] N. Ferreira, M. Iglesias, and F. Noronha, Granitoides da Zona Centro Ibérica e o seu enquadramento geodinâmico.*Geologia de los Granitoides y Rocas Asociadas del Macizo: Libro Homenage a L.C. Garcia de Figuerola.,* F. Bea, Ed., Editorial Rueda: Madrid, 1987, pp. 37-51.

[31] F. Noronha, Nota sobre a ocorrência de filões com espodumena na folha de Dornelas, *Portuguese Geological Survey, Internal report,* 1987, p. 2.

[32] B. Charoy, F. Lothe, Y. Dusausoy, and F. Noronha, "The crystal chemistry of spodumene in some granitic aplite-pegmatite from Northern Portugal", *Can. Mineral.,* vol. 30, pp. 639-651, 1992.

[33] B. Charoy, F. Noronha, and A. Lima, "Spodumene-Petalite-Eucryptite: mutual relationships and alteration style in pegmatite-aplite dykes from Northern Portugal", *Can. Mineral.,* vol. 39, pp. 729-746, 2001.
[http://dx.doi.org/10.2113/gscanmin.39.3.729]

[34] A. Lima, R. Vieira, T.C. Martins, F. Noronha, and B. Charoy, *A ocorrência de petalite como fase litinífera dominante em numerosos filões do campo aplitopegmatítico do Barroso-Alvão (Norte de Portugal) [CD-ROM].* VI Congresso Nacional de Geologia: Almada, 2003, pp. F52-F55.

[35] T. Martins, A. Lima, W.B. Simmons, A.U. Falster, and F. Noronha, "Geochemical fractionation of

Nb-Ta oxides in Li-bearing pegmatites from the Barroso-Alvão pegmatite field, Northern Portugal", *Can. Mineral.,* vol. 49, pp. 777-791, 2011.
[http://dx.doi.org/10.3749/canmin.49.3.777]

[36]   T. Martins, E. Roda-Robles, A. Lima, and P. Parseval, "Geochemistry and evolution of micas in the Barroso-Alvão pegmatite field, Northern Portugal", *Can. Mineral.,* vol. 50, pp. 1117-1129, 2012.
[http://dx.doi.org/10.3749/canmin.50.4.1117]

[37]   E.J. Carranza, "Controls on mineral deposit occurrence inferred from analysis of their spatial pattern and spatial association with geological features", *Ore Geol. Rev.,* vol. 35, pp. 383-400, 2009.
[http://dx.doi.org/10.1016/j.oregeorev.2009.01.001]

[38]   G.F. Bonham-Carter, *Geographic Information Systems for Geoscientists - Modelling with GIS. 1.* Pergamon: Ontario, 1994.

[39]   M. Vaillant, J.M. Jouany, and J. Devillers, "A multicriteria estimation of the environmental risk of chemicals with the SIRIS method", *Toxicol. Model.,* vol. 1, pp. 57-72, 1995.

[40]   E.J. Carranza, "Geocomputation of mineral exploration targets", *Comput. Geosci.,* vol. 37, pp. 1907-1916, 2011.
[http://dx.doi.org/10.1016/j.cageo.2011.11.009]

[41]   D. Cassard, M. Billa, and A. Lambert, "Gold predictivity mapping in French Guiana using an expert-guided data-driven approach based on a regional- scale GIS", *Ore Geol. Rev.,* vol. 34, pp. 471-500, 2008.
[http://dx.doi.org/10.1016/j.oregeorev.2008.06.001]

[42]   P.J. Clark, and F.C. Evans, "Distance to nearest neighbor as a measure of spatial relationship in population", *Ecology,* vol. 35, pp. 445-453, 1954.
[http://dx.doi.org/10.2307/1931034]

[43]   M. Berman, "Distance distributions associated with Poisson processes of geometric figures", *J. Appl. Probab.,* vol. 14, pp. 195-199, 1977.
[http://dx.doi.org/10.1017/S0021900200104796]

[44]   M. Berman, "Testing for spatial association between a point processes and another stochastic process", *Appl. Stat.,* vol. 35, pp. 54-62, 1986.
[http://dx.doi.org/10.2307/2347865]

[45]   BD Ripley, "Modelling spatial patterns", *Journal of the Royal Statistical Society,* vol. 39, no. Series B, pp. 172-192, 1977.

[46]   G. Matheron, "Traité de Geostatistique Appliquée", *Mémoire du Bureau de Recherche Géologique et Minières,* vol. 24, p. 333, 1962.

[47]   C. Gumiaux, D. Gapais, and J.P. Brun, "Geostatistics applied to best-fit interpolation of orientation data", *Tectonophysics,* vol. 376, pp. 241-259, 2003.
[http://dx.doi.org/10.1016/j.tecto.2003.08.008]

[48]   Granitic Pegmatites, *The state of the art - Field trip guidebook. Memórias Nº 9.* Universidade do Porto, Faculdade de Ciências, Departamento de Geologia: Porto, 2007.

[49]   Prospeg Project, *Pegmatites remote sensing and mapping. Final report.* Sinergeo, Lda. and Universidade of Minho, 2013.

[50]   R. Ramos, *"Condicionamento tectono-estratigráficos e litogeoquímicos da evolução metamórfica Varisca, nas unidades parautóctones (Trás-os-Montes Ocidental)",* PhD dissertation. Portugal: University of Porto, 2012.

[51]   P. Černý, Exploration strategy and methods for pegmatite deposits of tantalum.*Lanthanides, Tantalum and Niobium.,* P. Moller, P. Černý, F. Saupe, Eds., Springer-Verlag: Berlin, 1989, pp. 274-302.
[http://dx.doi.org/10.1007/978-3-642-87262-4_13]

[52]   D. London, "A petrologic assessment of internal zonation in granitic pegmatites", *Lithos,* vol. 184-187, pp. 74-104, 2014.
[http://dx.doi.org/10.1016/j.lithos.2013.10.025]

Frontiers in Information Systems, 2018, *Vol. 1*, 102-123

CHAPTER 4

Open Source GIS Tools: Two Environmental Applications

Lia Bárbara Cunha Barata Duarte[1,2,*], **José Alberto Álvares Pereira Gonçalves**[3] and **Ana Cláudia Teodoro**[1,2]

[1] *Department of Geosciences, Environment and Land Planning, Faculty of Sciences, University of Porto, Porto, Portugal*

[2] *Earth Sciences Institute (ICT), Faculty of Sciences, University of Porto, Porto, Portugal*

[3] *Interdisciplinary Centre of Marine and Environmental Research (CIIMAR), University of Porto, Porto, Portugal*

Abstract: Geographical information systems (GIS) incorporate robust tools that allow for the incorporation, manipulation and analysis of different types of data, such as geologic, hydrogeological, meteorological and environmental, and to display large amounts of geocoded data, useful to map the spatial distribution of natural phenomena. Groundwater pollution and soil erosion are some of the environmental concerns at global scale, that require efficient mapping tools. An assessment of the groundwater vulnerability and soil loss through open source applications, developed for this purpose, is a valuable contribution to several communities. The applications presented in this work were developed within the QGIS software, using several open source libraries. The first application was developed to produce maps to evaluate the groundwater vulnerability to pollution. The tool integrates the procedures required to implement the DRASTIC index under a single plugin. The application is easy to use and provides the possibility of importing the attribute table, and allows for the possibility of modifying weights, indexes and attributes, in an interactive manner. Maps can be generated according to the user perception, regarding each aquifer system. The second application is intended to estimate the expected soil loss by water-caused erosion, using the Revised Universal Soil Loss Equation (RUSLE) through a web browser. This application provides the tools to manipulate the input data of the RUSLE model and to create categorical maps needed to assess the risk of soil loss. This web application was implemented in order to be used by users without GIS software skills. In order to test the two applications developed, two study cases were performed: in River Zêzere Basin Upstream of Manteigas (Serra da Estrela) for DRASTIC index and Montalegre municipality for RUSLE. The resulting maps met the expectation of soil scientists for these study areas.

---

* **Corresponding author Lia Bárbara Cunha Barata Duarte:** Department of Geosciences, Environment and Land Planning, Faculty of Sciences, University of Porto, Porto, Portugal; Tel: 913687661; E-mail: liaduarte@fc.up.pt

Keywords: DRASTIC, Environment impacts, GIS, Groundwater vulnerability, Open source, Python, QGIS, RUSLE, Soil erosion, Web.

INTRODUCTION

Open Source Software

Geographic Information System (GIS) technology is used to manipulate and analyse geospatial data, and also as a decision support system [1,2]. Many fields of study use this technology, but it is particularly important in environmental management. The implementation of a GIS comprises several components, such as the appropriate spatial data, and obviously the software tools to manage, analyse and display the data [3]. Most of the GIS implementations have been based on proprietary software. In recent years, open-source based software had a significant improvement in the field of GIS and some became very popular. One reason is the cost, since most of it is free. However, of not less importance is the fact that, due to being developed by a community that shares knowledge and experience, many benefits come from the fact that new modules can be implemented, based on easy to use programming tools that expand much the capability of GIS data analysis. The open source concept follows the four freedoms: freedom to run the program by anyone with different objectives, to study the program code and modify it, to provide copies and freedom to provide modified copies of the code [4]. QGIS desktop software is free and is one of the most popular open source GIS software [2]. Gary Sherman started QGIS in 2002 [5]. It is licensed under a GNU General Public License (GPL) license. It is developed in C++ programming language providing a good online maintenance related with the plugin development and the use of the program. Plugins (known as extensions or tools) contribute to software expansion [6]. QGIS plugins are developed in C++ or Python language.

Python is a programming language easy to use, simple and rapid in the application development. It is used in several applications domains and is similar to other languages (JAVA, Tcl, Perl, Scheme or Ruby). It is open source and runs in Windows, Linux/ Unix, OS2, Mac and other operating systems. There is a lot of documentation and tutorials online. To create a new plugin, QGIS provide a specific procedure [5].

1. Idea: Define the purpose of the new plugin.
2. Create the required files: Follow the rules defined by QGIS.
3. Write the main code: Write the code.
4. Do some tests: Do some tests during the developing phase.
5. Publish the plugin: In the end, the plugin can be available in the QGIS official repository or through a personal repository.

Several files must be created and a directory structure must be followed. __init__.py script is the initial point of the plugin and plugin.py script presents the main code. A plugin is composed by several scripts with code containing all the information about the plugin actions and configurations. The scripts with form.ui are related to the application interface: windows, edit lines, combo boxes, buttons and other interface options. The scripts with form.py presents the main code associated to form.ui.

QGIS contains Application Programming Interfaces (APIs): QGIS API, GDAL/OGR API and PyQt4 API. These APIs are composed by functions, classes and modules. These modules allow to interact with geographic information. QGIS API and PyQt4 API are composed by Core and GUI modules and GDAL/OGR libraries allow the manipulation of raster files and vector information. QGIS provides the connection with several open source libraries such as *Numpy* library [7], *matplotlib* library [8], and SEXTANTE (*Processing Toolbox* more recently), a spatial data processing framework. SEXTANTE provides spatial analysis and functionalities allowing to manipulate geographic data into QGIS [5]. Also, another advantage from QGIS is the possibility of supporting external applications (GRASS, SAGA or Orfeo Toolbox) with very useful algorithms which are integrated in the *Processing Toolbox* framework from QGIS. SEXTANTE is also free with GPL2 license [5, 7, 9 - 14].

GIS and Environmental Issues

Many factors affect the environment causing negative impacts and serious problems in several areas. Groundwater pollution and soil erosion are two of these factors. There is a relation between these two factors, because groundwater contamination is related to the use of agricultural chemicals and reflects the propensity of soils to leach pesticides and nitrates.

Groundwater vulnerability and soil erosion are often estimated with GIS which provides the required tools to manipulate, analyse and incorporate geographic data, such as geologic and hydrogeological information [15]. GIS have been used to assess groundwater vulnerability based on DRASTIC index in several studies [16 - 19] or modified DRASTIC methods [20, 21].The groundwater vulnerability and risk of pollution are also determined by other methods [22]. GIS proprietary software, such as ArcGIS, was mainly used in these studies [23 - 26]. Also, RUSLE method is used to assess soil erosion through GIS [27 - 32]. As in DRASTIC, all the studies conducted considering the RUSLE method were performed in proprietary software. So, the user must have/pay a license to use. Moreover, it is not so flexible to adapt the code to other study cases, as it is not possible to modify the source code. Thereby, applications based on open source software are an added value for the models referred, since the source code is available and any user can modify. Also, the code can be adapted to other regions.

The objective of the work presented in this chapter was the creation of open source applications with different versions, desktop and web, to create the maps involved in RUSLE method (web application) and to assess groundwater vulnerability to pollution through the DRASTIC index (desktop application). The web page was developed using DJANGO framework [33]. This page returns the maps based on RUSLE method, without having GIS skills and the QGIS software installed. DJANGO is a free and open source project, a high level Python Web application framework designed to support the development of dynamic websites and web applications. It is developed in Python and follows the Model-View-Controller (MVC) architectural pattern. It is sustained by a non-profit independent organization, the Django Software Foundation (DSF) [33]. The two versions developed are free. The desktop application (DRASTIC) can be installed in QGIS and the web application (RUSLE) can be accessed through a web browser. The open source application presents the advantage of having simple and intuitive graphic interfaces, allowing to create maps and spatial analysis under a desktop environment. Being based on open source code they have the advantage of being easy to adapt to other possible needs of users.

METHODOLOGY

Groundwater is a natural resource of high economic value. The demand for groundwater is growing every year due to the pollution of surface water resources [34, 35]. Foster [36] defines aquifer pollution vulnerability as the intrinsic characteristics to determine the sensitivity of the groundwater system to being adversely affected by an imposed pollutant load. The probability for pollutants to reach a position in the groundwater system when introduced at some location above the uppermost aquifer defines the groundwater vulnerability to pollution [37]. Assessment and mapping of groundwater vulnerability is extremely important for aquifer management. Many types of factors are involved in aquifers: geological, geomorphological, climatic and biological conditions, landforms, unsaturated zone, aquifer hydraulic features and land use. The complex process requires different models with different ratings, statistical methods, process-based methods and index methods [15]. One of the most used methods is DRASTIC index applied as the original index developed by Aller *et al.* [38] or using modified versions [*e.g.*, 24, 39, 40].

Also, a severe and major environmental problem is the soil erosion, which is the detachment of individual soil particles from the soil mass transported by erosive gents [43]. This phenomenon endangers agriculture, environment, natural resources and contributes to flooding and habitat destruction, and affects the water reservoirs [41, 42]. The economic and environmental consequences from the soil erosion are difficult to quantify [44]. Consequently, it is crucial to define policies and prevention measures on water and soil resource conservation [41]. Several factors affect and interfere with soil erosion, such as soil degradation, water quality, hydrogeological systems, agricultural production, vegetation cover, topographic features, climatic variables, soil, slope and the most important, the precipitation. According to Nearing *et al.* [45], increasing precipitation intensity, increases erosion; decreasing precipitation, due to the interactions with plants biomass and runoff, erosion can either increase or decrease. Thereby the precipitation intensity and precipitation amount is important due to climate change impacts on soil erosion [45]. Even though soil erosion is defined as a natural phenomenon, the human impacts and agriculture activities change the vegetation cover and accelerate the problem [42]. Currently there are several methods to estimate the soil loss and assess the soil erosion risk, such as Universal Soil Loss Equation (USLE) [46], Revised Universal Soil Loss Equation (RUSLE) [47], Revised Universal Soil Loss Equation-3D (RUSLE-3D) [60] and Revised Universal Soil Loss Equation embedded with the Information Diffusion Model (RUSLE-IDM) [41], among others. RUSLE method uses the same equation as USLE, an empirical model, but includes improvements in factors calculation [48] and the Water Erosion Prediction Project (WEPP) [49], topographic factor (*LS*) and the product of slope length (*L*) with slope steepness (*S*). The RUSLE was proposed as a conservation plan and assessment tool in 2005 by the USDA-Agricultural Research Service [49]. The RUSLE is used to control soil erosion developing conservation plans. RUSLE integrates factors that affect soil erosion such as climate, topography and other, which are easily available [42]. However, when the data is not completely available, several procedures must be optimized due to the time processing.

DRASTIC Index

The DRASTIC index, according to Aller *et al.* [38], is composed by seven factors: Depth to Groundwater (*D*), Net Recharge (*R*), Aquifer Media (*A*), Soil Media (*S*), Topography (*T*), Impact of the Vadose Zone (*I*) and Hydraulic Conductivity (*C*) where each factor is multiplied by the weight and in the end all of the maps are summed as defined in eq. (1).

$$DRASTIC = D_R \times D_W + R_R \times R_W + A_R \times A_W + S_R \times S_W + \\ T_R \times T_W + I_R \times I_W + C_R \times C_W \tag{1}$$

Where *R* and *W* are the rating and weight for each factor, respectively.

DRASTIC Desktop Application

The DRASTIC application is composed by a single window (Fig. **1**). This window is composed by a map canvas and a menu bar with *File* menu, the *DRASTIC* menu and the *Help* menu. The *File* menu contains buttons that allow the user to add a vector or a raster file (*Add Vector Layer- F1* and *Add Raster Layer-F2*). The *DRASTIC* menu has seven buttons (F1 to F7), one for each factor and a last one to DRASTIC index (F8).

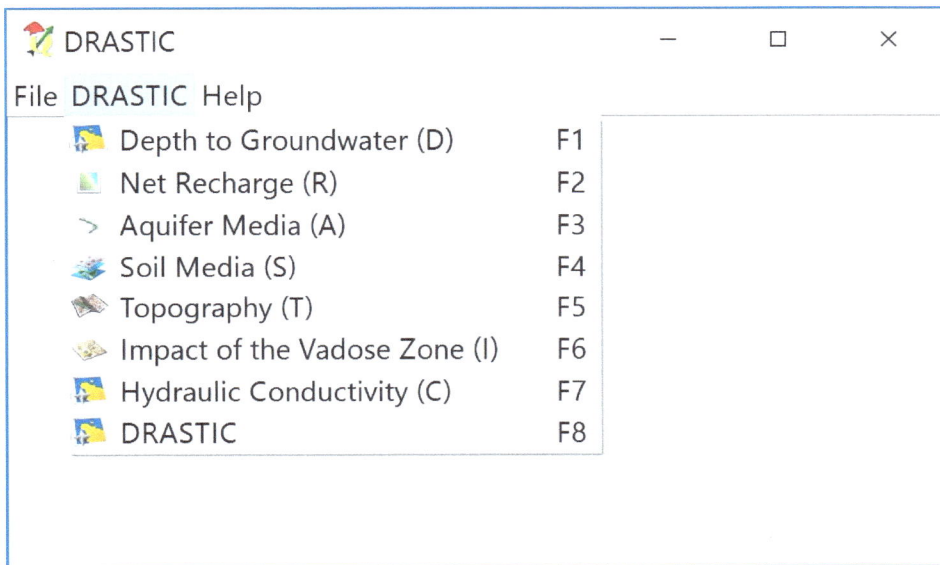

Fig. (1). DRASTIC graphic interface [13].

The last button, *DRASTIC*, allows to create the groundwater vulnerability map using the seven raster files previously generated, according to the DRASTIC index expression from Aller *et al.* [38], presented in eq. (1). Some information about the tool and the factors is provided in the *Help* menu (Fig. **2**).

The graphic interfaces include: an input field, an attribute field, a cell size definition, a weight definition, the table for ratings definition and a field to output directory. In all factors, the user has the flexibility to modify the ratings in the attribute table and add/remove classes. The modification of the weight values is also possible.

The *A*, *S* and *I* factors provide by default the model adopted by Aller *et al.* [38]. The users can modify the ratings or the descriptions. The button *Attribute Table* allows to import the input attribute table.

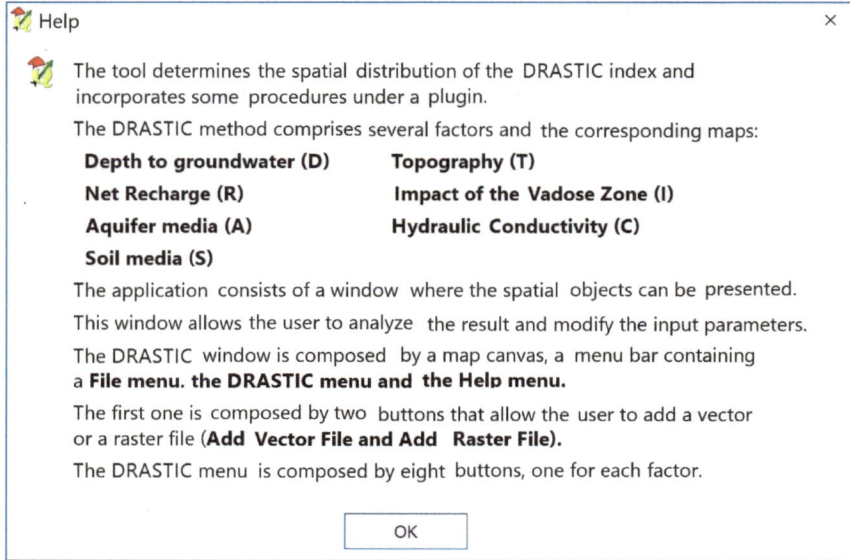

Fig. (2). DRASTIC *Help* button [14].

DRASTIC Factors

D factor allows to control the distance that pollutants travel before reaching the aquifer. A surface map based on depth values measured in the wells is created. Under the application, the surface can be created using two methods: the base method and an improved method that uses a DEM (Digital Elevation Model). To estimate the map, different interpolation methods are available. The improved method defines the DEM as input raster file and it is based on drainage network segments. More information can be obtained in Duarte *et al.* [14].

The greater the aquifer recharge, the greater is the groundwater vulnerability to pollution. To estimate the *R* factor map three methods are provided. Depending on the available information, the user chooses the most adequate method. The first method estimates the factor using a simplified water budget (*e.g.,* [50, 51]), given by eq. (2):

$$Recharge = Precipitation - Overland\ Flow - Evapotranspiration \qquad (2)$$

The second method determines the recharge as a percentage of mean annual precipitation data (mm/year). The precipitation is controlled by altitude, so a third method was incorporated calculating the precipitation based on DEM with a regression model [14].

The *A* factor is based on the available geological information regarding the influence of the geologic material on the groundwater vulnerability to pollution. The application reads and analyzes the input attribute table and creates a new column with the new ratings.

The *S* factor corresponds to the influence of soil thickness and texture on the attenuation of the pollution. This data can be obtained from soil maps or other sources and the feature is identical to *A* feature.

The *T* factor corresponds to the influence of terrain surface slope on the infiltration of polluted water into the soil. In the developed application two different methods were incorporated: the feature creates the DEM if a contour shapefile is available, derives the slope and reclassifies according to the ratings; if the user already has the DEM, it is specified as input file and the slope and reclassification is performed.

The *I* factor concerns the control of pollutant attenuation when a specific geologic material is located below the soil layer and above the saturated zone. This information is created as *A* factor.

The *C* factor relies on the fact that, for a given aquifer material, it is proportional to the groundwater vulnerability to pollution. Hydraulic conductivity values can be available in a geological map and are based on field tests. If there is a lack of data, typical values for hydrogeological conditions can be adopted.

The mathematical multiplication of the seven factors, affected by weights, referred above allows to obtain the DRASTIC index map. The user must specify the factors as raster files, the weight assigned to each factor and the output file directory [14].

RUSLE Model

The RUSLE model involves six erosion factors to estimate the mean annual soil loss, as presented in eq. (3):

$$A = R \times K \times LS \times C \times P \qquad\qquad (3)$$

Where *A* corresponds to the average annual soil loss (ton km^{-2} $year^{-1}$), *R* is the erosivity factor (MJ mm km^{-2} h^{-1} $year^{-1}$), *K* is the soil erodibility factor (t km^{2} h km^{-2} MJ^{-1} mm^{-1}), *LS* is the combination of *L* and slope *S* (dimensionless), *C* is the land cover factor (dimensionless) and *P* is the practices used for erosion control (dimensionless, ranging between 0 and 1).

RUSLE Web Application

The RUSLE web application was created based on several scripts. Fig. (3) shows a diagram of the application structure.

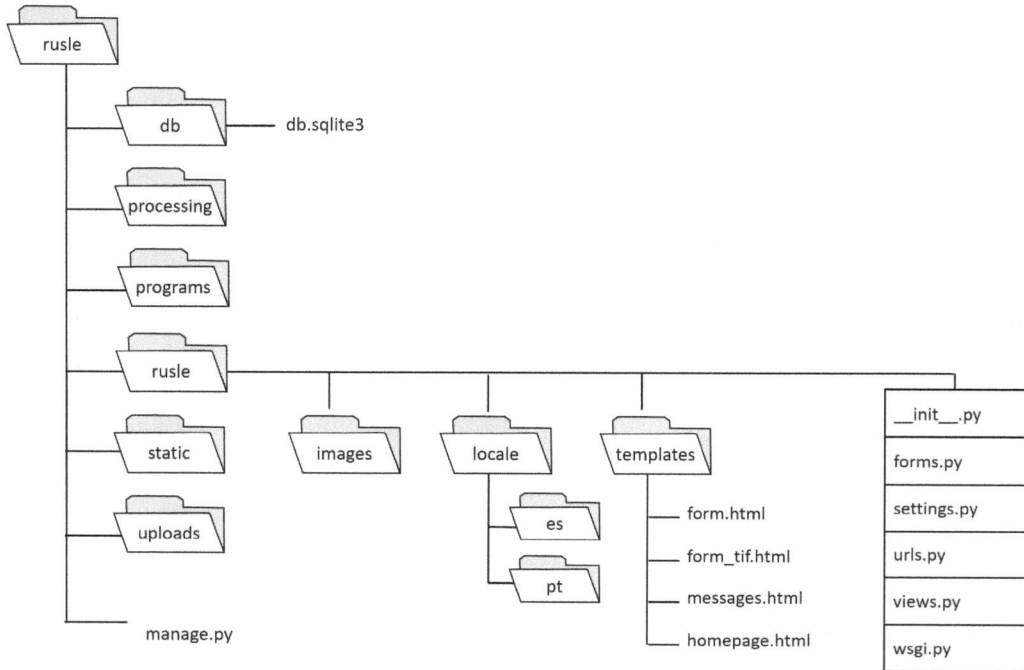

Fig. (3). Diagram of the application file structure.

The folders created include the QGIS algorithms used, the Python packages of the website and the HTML templates used. In the *rusle/* folder, several scripts were created in order to define the project configurations and the URL declarations, the manipulation between the updated files, the creation of the maps and the return of the resulting map. If some definition is incorrect, warning messages are displayed.

The web page contains *check boxes*, *combo boxes* and other types of widgets. The input elements are specified in a form (field to input directory) which was built in HTML language. To help the user, a *help* text was created to each field. The front page was defined with the RUSLE method description and the parameters definition. The page is composed by a header, the navigation links in the lateral tab and the footer, where the user can email the author/authors and the body. The lateral tab is defined with a list of all the factors incorporated in the RUSLE map, as shown in Fig. (4).

RUSLE (Revised Universal Soil Loss Equation)

R Factor

K Factor

LS Factor

C Factor

P Factor

RUSLE Factor

Soil Erosion Risk Maps

Soil erosion is a serious environmental problem worldwide. An approximation of the expected soil loss by water-caused erosion can be calculated by with the Re-vised Universal Soil Loss Equation (RUSLE). GIS (Geographical Information Systems) provide different tools to create categorical maps of soil erosion risk which helps to study the risk assessment of soil loss.

RUSLE model incorporates six erosion factors to understand the mean annual soil loss:

A=RKLSCP

where A is the computed estimation of average annual soil loss (t km-2 year-1), R is the rainfall-runoff erosivity factor (MJ mm km-2 h-1 year-1), K is the soil erodibility factor (t km2 h km-2 MJ-1 mm-1), LS is the combination of slope length (L) and slope steep-ness (S) (dimentionless), C is the land cover factor (dimensionless) and P is the prac-tices used for erosion control (dimensionless, ranging between 0 and 1).

Rainfall erosivity factor (R)

Wischmeier (1978) defines rainfall runoff erosivity factor as the factor that indi-cates the soil loss potential of a given storm event and represents the effect of rainfall intensity on soil erosion. When

Fig. (4). RUSLE Web page.

The resulting map is displayed in the page when the user assigns the input files. The message "x files uploaded. Map generated: XFactor.jpg." is displayed when the input files are correctly inserted. The x is the number of files updated and XFactor is the name of the map created. The final RUSLE map is composed based on the files already created and stored in the application. The maps are created in JPG format and downloaded using *Download* option. A layout with the fundamental elements is created through *QgsComposerMap* class from QGIS API. The legend is defined with a color ramp from low (green) to high (red) values.

RUSLE Factors

R factor represents the effect of rainfall intensity on soil erosion [46]. However, the soil erosion could be estimated by the erosion index (EI30) when other factors are constant. The EI30 is the sum of the kinetic energy (MJ/ha), multiplied by the greatest amount of rain in any 30 minute period (mm/h) [52]. Due to the unavailable EI30 value, factor R can be estimated from several algorithms. In this application, the R factor only incorporates one method, rainfall derived EI30, according to Loureiro and Coutinho [53], which can be obtained from meteorological data. The points file is interpolated through the Inverse Distance Weighting (IDW) interpolation method [54]. In order to define the study area extents, a shapefile input mask is required.

K factor represents the susceptibility of a soil type to erosion per unit erosivity of rainfall relating soil texture, organic matter and permeability [46, 47]. This factor is calculated through the algorithm proposed by Wischmeier *et al.* [52]:

$$K = \frac{2.1 \cdot 10^{-4} \cdot (12 - OM) \cdot M^{1.14} + 3.25 \cdot (s - 2) + 2.5 \cdot (p - 3)}{100} \tag{4}$$

In eq. (4), *OM* corresponds to the organic matter (%), *s* to the soil structure, *p* to the permeability and *M* corresponds to the aggregated variable derived from the granular soil texture which can be estimated from: *M=(%Msilt)×(%silt+%sand)* where the modified silt (*Msilt*) is a percentage of grain size between 0.002 and 0.1 mm. In the application the input field is, usually, the soil map shapefile. To simplify the dispersed, varied and heterogeneous data, several tables composed by erodibility values based on SROA (Serviço de Reconhecimento e Ordenamento Agrário) were incorporated [55]. The tables were defined for Portugal, making the application adapted to the characteristics of the country. However it can be modified for other countries rules. The web application automatizes the process, reading the attribute table and listing the values from the same parameter to the Portugal tables. To obtain the *K* value, the eq. (4) is computed for each soil type. Finally, *K* values were used as input to convert the shapefile to raster format using GRASS *v.to.rast.attribute* algorithm.

LS factor is related with the topography effect in erosion [47]. The combined *LS* factor in RUSLE represents the ratio of soil loss on a given *L* and *S* to soil loss from a slope that has a length of 22.6 m and a steepness of 9% [47]. These factors are typically evaluated together and could be obtained from contour maps or DEM or directly in the field [47]. From the several existing equations to estimate the *LS* factor, the equation proposed by Wischmeier and Smith [46] was incorporated in the application. This equation uses *S* and *L* as a single index, expressing the ratio of soil loss, as shown in eq. (5):

$$LS = \left(\frac{X}{22.1}\right)^m (0.065 + 0.0456 \times S + 0.0065 \times S^2) \tag{5}$$

In this equation, *X* corresponds to slope length (meters) and *S* corresponds to slope steepness (in percentage). These values can be obtained from a DEM. The *m* value depends of the slope percentage, and ranges from 0.2 to 0.5, as shown in Table **1**.

Table 1. Relation between slope and *m* value.

Slope (%)	*m*
≥5%	0.5
3.5-4.5	0.4
1-3	0.3
<1% or uniform gradients	0.2

The *LS* factor implementation on RUSLE considers several algorithms from *Processing Toolbox*. The *S* corresponds to the slope created from a DEM and in the application this factor is obtained using algorithm *r.slope.aspect*, from GRASS. The *L* (in meters) is created through *slope length* algorithm from SAGA. To obtain the *LS* raster factor, *raster calculator* algorithm, from SAGA, and eq. (5) were used.

The *C* factor is used to understand the existing crops and management practices effects on erosion. For bare soil the *C* value has value 1 and decreases as the soil cover increases, implying less soil erosion. Different approaches have been used to estimate this factor such as image classification techniques [41, 42, 54, 56], and CLC (Corine Land Cover) maps [57, 58]. Based on that, several tables were developed for Portugal and published [55].

In the application, CLC map or the COS (Carta de Ocupação do Solo) data can be used as input to calculate the *C* factor. In the web application, the uploaded shapefile must contain the assigned *C* factor values (0 to 1 scale) to each land use class. Finally, through *v.to.rast.attribute* algorithm, the shapefile is converted to raster format.

P factor (ranging from 0 to 1) incorporates the effects of conservation practices to protect the soil [47]. When the adopted conservation practice reduces soil erosion, *P* factor assumes values smaller than 1 and when the land is directly cultivated on the slope, *P* factor values are equal to 1.

Kumar and Kushwaha [58] proposed a method based in Land Use/Land Cover (LULC) information to estimate the *P* factor. Considering the different types of LULC, the *P* factor values are defined according to the literature as cited in [58]. These values must be assigned before uploading the file into the web application. To convert the shapefile with the *P* values assigned to raster format an algorithm from GRASS is used.

RESULTS

In order to test the two applications developed, they were applied and evaluated in two case studies (Fig. **5**). DRASTIC methodology was tested in a particular region of Serra da Estrela (Guarda, Centre part of Portugal). RUSLE web application was tested in the Montalegre municipality (Vila Real, North part of Portugal).

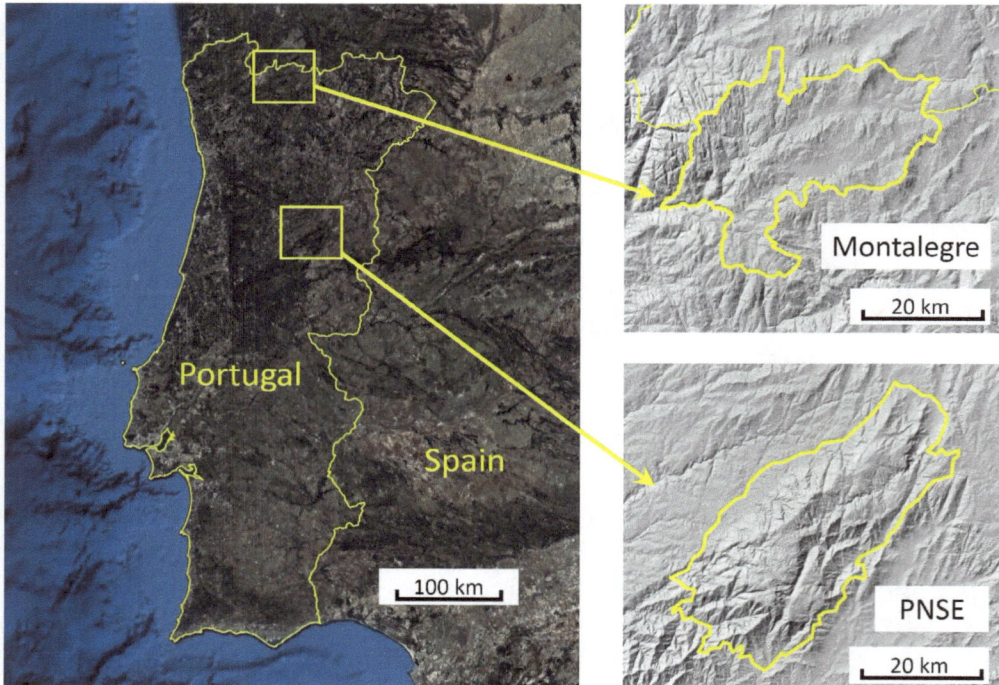

Fig. (5). Location of case studies.

DRASTIC Study Case

The DRASTIC study case is located in the River Zêzere Basin Upstream of Manteigas village (ZBUM), in Serra da Estrela region (40°19'N, 7°37'W), Central Portugal. This study area presents a complex hydrogeological system and it can prove how the application is capable of handling the assessment of groundwater vulnerability [60]. To understand better the water cycle in a mountain environment, several hydrogeological studies have been developed in ZBUM region [59 - 61]. Several factors contribute to control the water cycle such as climatic, geologic and geomorphologic features [61]. The assessment of groundwater vulnerability to pollution method based on DRASTIC index provides a tool to improve the sustainability of the water resources management in Serra da Estrela.

The *D* factor and the *R* factor maps have been created using the DEM. In the region and according to the field observations, the water table depth is at least 20 meters in hilltop landform areas, whereas it reaches the topographic surface under river and stream valley bottoms. The *R* factor map was created from the DEM, estimating the spatial distribution of mean annual precipitation, using the regression model *y = 0.99x + 542.22*, where *y* is the mean annual precipitation (in mm) and *x* is the altitude (in m) [60] and applying the regional aquifer recharge rate of 15% [60]. The *A*, *S* and *I* factor maps were created using the geological map of Serra da Estrela Natural Park at scale 1/75 000 from [62] as well as hydropedological information [60, 61, 63]. Geological material was rated according to Aller *et al*. [38] to estimate *A* and *I* factor maps and the soil texture classes presented by Espinha Marques *et al*. [63] were used to estimate *S* factor map that is, mostly sandy soils. The ratings were assigned according Aller *et al*. [38] in the imported attribute table. A DEM from topographic maps (1/25 000) was used to obtain factor *T*. For factor *C*, the hydraulic conductivity value of 10^{-4} cm/s was assigned to igneous and metamorphic rocks and 10^{-2} cm/s was assigned to glacial deposits. In the end, the final map based on DRASTIC index was created (Fig. **6**).

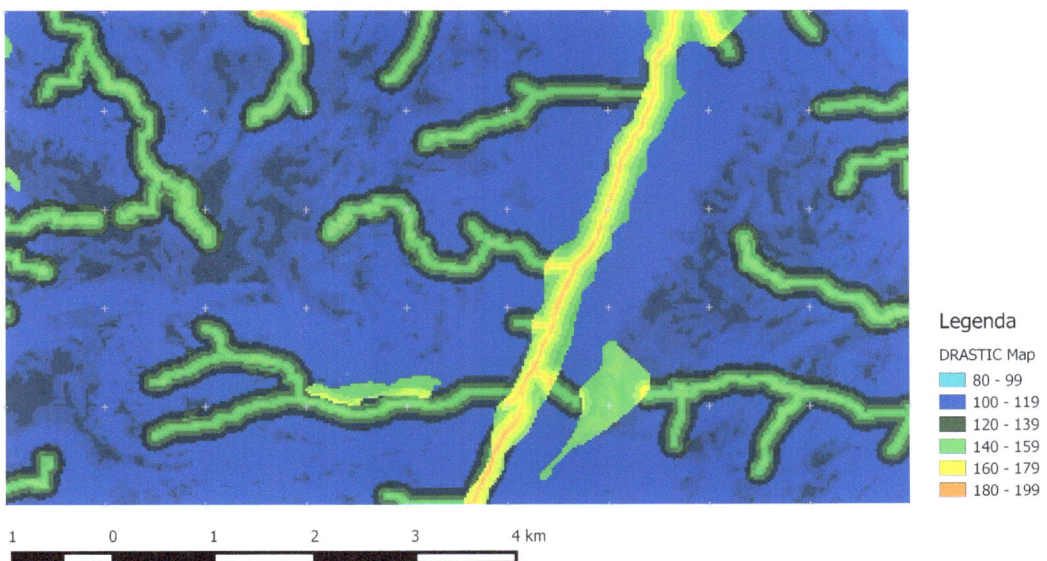

Fig. (6). DRASTIC index map [14].

RUSLE Study Case

The RUSLE web application was tested for the Montalegre municipality, in North of Portugal, covering 80 500 hectares. This region is characterized by high slopes leading to high erosion risks when soil protection measures are not taken. This study case was preferred because there is available data and due to the high slopes and different LULC classes in the region. The spatial resolution of the raster files used was 20 m and the coordinate system adopted was ETRS89/PT-TM06 (EPSG:3763). To use the web application, the factor values must be previously defined by the user in the attribute table of each shapefile.

The R factor map was obtained using meteorological data from Montalegre meteorological station. A monthly rainfall database (January 1981 to December 2008 (28 years)) was considered in this study. The monthly rainfall was estimated for the days with more than 10.0 mm (*rain10*) and the monthly number of days with rainfall higher than 10.0 mm (*days10*) for each year. Through the station location, the web application creates the R factor raster through the IDW interpolation method. The R value is the sum of erosive storm EI30 values occurring during a year which is computed in the application according to Loureiro and Coutinho [53] through *raster calculator* algorithm from SAGA. In this study case there was only one station available, so the R factor raster is defined as a constant value. The K factor was obtained from the Montalegre soil map. The values were defined previously according to the tables for Portugal [55] which adjust the available alphanumeric and cartographic information related to LULC and soil types [55]. The C factor and the P factor were obtained from COS2007 (Carta de Ocupação do Solo 2007, level 2). Also in this factor, the values were assigned based on the Portugal tables [55]. To map the soil erosion risk areas and correctly interpret the map, a set of classes were defined: very high, high, medium, low and very low. Fig. (7) presents the RUSLE map result for Montalegre municipality.

DISCUSSION

The DRASTIC study case was the hydrogeological system located in the ZBUM area. In some factors, namely, D, R, T and C, the ratings adopted were based on the literature. In the other factors, the ratings were adopted according to the experts' knowledge on the system's hydrogeological conditions. The final DRASTIC map was created using the weights defined by Aller *et al.* [38]. In the results obtained the majority of the study area (19,4%) was characterized with moderate vulnerability values (120 to 159), while the higher vulnerability values (160 to 181) correspond to 1,6% and the lower vulnerability values (90 to 119) correspond to 79% of the area. Through Fig. (6), it was concluded that the higher vulnerability zones correspond to valley bottom landforms. The lower vulnerability values are located in hillsides, in zones where the water table is deeper, where there are granitic or metamorphic rocks and higher slopes. The DRASTIC map presents different degrees of vulnerability which is a consequence of the hydrogeological system of the study area.

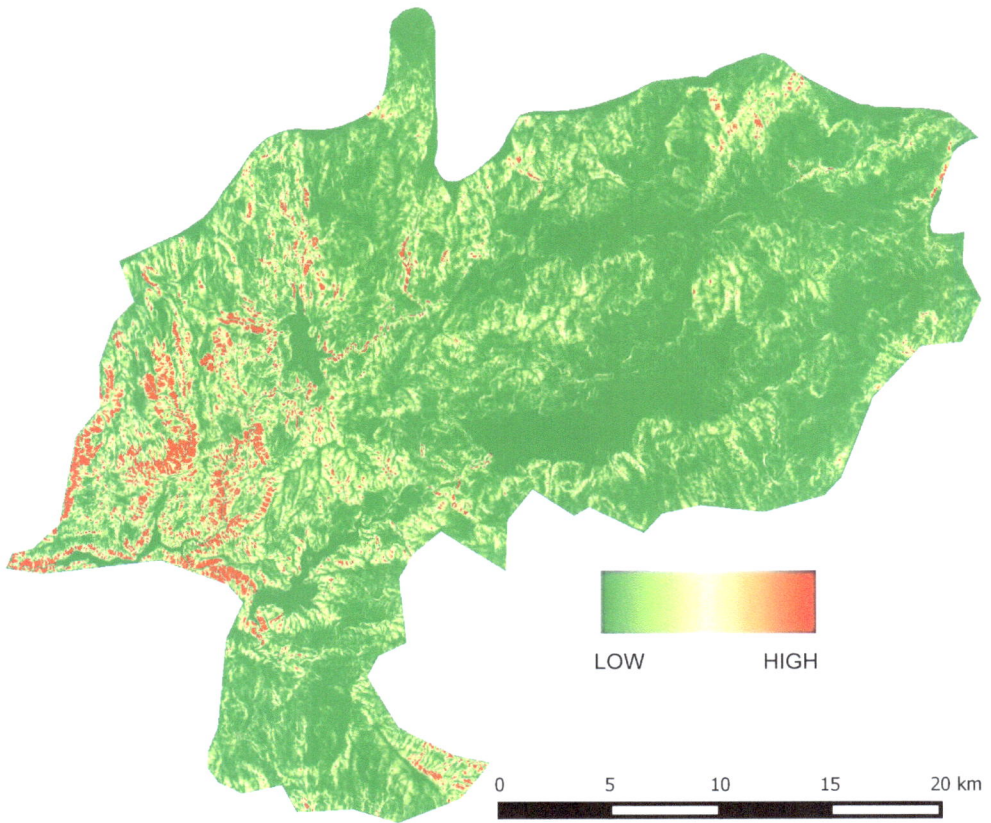

Fig. (7). Soil erosion map.

The RUSLE web application was tested for Montalegre municipality. A moderate erosion risk was suggested in RUSLE map in West side of Montalegre and a low erosion risk was identified in the East side of the municipality. In the final map it was also identified 29% of the area as very low soil erosion level, only 6% was classified as high and 8% as very high soil erosion level. Very high slope values were identified in 7% of the total area. In the final risk map, 4% of the total area was characterized as very high soil erosion combined with high slope values. Montalegre municipality has typical high values of precipitation and high slope values; however most of the area was classified with low levels of risk. This can be explained by the soil protection with permanents crops. The data available for this study case were not enough to correctly evaluate the RUSLE factor. However, using the data available for the study zone, the final RUSLE factor map was obtained using the ratings presented in the tables for Portugal [55] and the incorporated model defined by Wischmeier and Smith [46].

CONCLUSIONS

The applications presented, one in a desktop GIS environment and the other as a web browser application, have several advantages in order to automate in a simple way the methods, functionalities and procedures incorporated. The main advantage of the DRASTIC desktop application is the possibility of modifying ratings, weights and methods to control the factors and the possibility of import the attribute table to easily assign the ratings. Other advantage is the fact that the application was developed as a single window where the results are immediately shown. Given that, the user evaluates and analyzes any resulting map interacting with the input parameters and creating another map if the results obtained are not the expected. Several methods were incorporated in the application, for instance D and R factors provide more than one method to obtain the result according to the available data. Help buttons were also incorporated in the application with descriptions of the required information about each feature and input/output information, so the user can easily create the maps. The application is open source so it is available (*http://www.fc.up.pt/pessoas/liaduarte/DRASTIC.rar*). It is easy to install and use and any user can modify the code or the methods in order to adapt to a different aquifer system. It is intended that the developed application can be very useful in hydrogeology community.

Soil erosion risk is also a severe environmental problem worldwide. Therefore, the creation of risk maps to evaluate the risks of soil loss is crucial. It was developed an open source web application to evaluate and study the risk assessment of soil loss implementing the best management practices. It is easy to use and provides help messages to guide the user. The main disadvantage is the fact that the user must define the ratings previously in the attribute table of the inputs. The user just needs to know the values of each case study and upload the data in a *zip* folder. As DRASTIC, the methods implemented in the application can be improved or modified with different rules, parameters or ratings. The web application has the advantage of not being necessary to have a GIS software installed and the user does not require the knowledge of tools and algorithms from any GIS software. The user just needs to know the values of each case study and upload the data. The web page is available, with a preliminary version currently accessible at http://www.geonet.fc.up.pt/rusle.

CONSENT FOR PUBLICATION

Not applicable.

CONFLICT OF INTEREST

The author (editor) declares no conflict of interest, financial or otherwise.

ACKNOWLEDGEMENTS

The authors would like to thank to Prof. Jorge Espinha, Prof. Mário Cunha and Prof. António Guerner for their collaboration and to Diana Soares for her help in the RUSLE web page creation.

REFERENCES

[1]   H. Mitasova, J. Hofierka, M. Zlocha, and L.R. Iverson, "Modelling topographic potential for erosion and deposition using GIS", *International Journal of Information Systems,* vol. 10, pp. 629-641, 1996.

[2]   A. Graser, and V. Olaya, "Processing: A Python Framework for the Seamless Integration of Geoprocessing Tools in QGIS", *ISPRS Int. J. Geoinf.,* vol. 4, pp. 2219-2245, 2015. [http://dx.doi.org/10.3390/ijgi4042219]

[3]   L. Larsen, GIS in environmental modeling and assessment, *Geographic Information Systems.,* P. Longley, M. Goodchild, D. Maguire, D. Rhind, Eds., 2nd ed. Wiley: New York, 1999, pp. 999-1007.

[4]   gnu.org [homepage on the Internet], *GNU Operating System..* http://www.gnu.org/philosophy/-philosophy.en.html

[5]   qgis.org [homepage on the Internet], *QGIS A Free and Open Source Geographic Information System..* http://www.qgis.org/en/site/

[6]   L. Duarte, and A.C. Teodoro, "An easy, accurate and efficient procedure to create Forest Fire Risk Maps using Modeler (SEXTANTE plugin)", *J. For. Res.,* vol. 27, no. 6, pp. 1361-1372, 2016. [http://dx.doi.org/10.1007/s11676-016-0267-5]

[7]   scipy.org [homepage on the Internet], *Numpy Reference.* http://docs.scipy.org/doc/numpy/reference/

[8]   matplotlib.org [homepage on the Internet], *Matplotlib Documentation.* http://matplotlib.org/

[9]   gdal.org [homepage on the Internet], *GDAL - Geospatial Data Abstraction Library.* http://www.gdal.org/

[10]   pyqt.sourceforge.net   [homepage   on   the   Internet],   *PyQt   Class   Reference.* http://pyqt.sourceforge.net/Docs/PyQt4/classes.html

[11]   saga-gis.org [homepage on the Internet], *SAGA System for Automated Geoscientific Analyses.* http://www.saga-gis.org/

[12]   osgeo.org [homepage on the Internet], *GRASS GIS Bringing advanced geospatial technologies to the world.* http://grass.osgeo.org/

[13]   A.C. Teodoro, and L. Duarte, "Forest Fire risk maps: a GIS open source application – a case study in Norwest of Portugal", *Int. J. Geogr. Inf. Sci.,* vol. 27, no. 4, pp. 699-720, 2013. [http://dx.doi.org/10.1080/13658816.2012.721554]

[14]   L. Duarte, A.C. Teodoro, J.A. Gonçalves, A.J. Guerner Dias, and J. Espinha Marques, "A dynamic map application for the assessment of groundwater vulnerability to pollution", *Environ. Earth Sci.,* vol. 74, no. 3, pp. 2315-2327, 2015.

[http://dx.doi.org/10.1007/s12665-015-4222-0]

[15] S.M. Shirazi, and H.M. Imran, "Shatirah Akib. GIS-Based DRASTIC method for groundwater vulnerability assessment: a review", *J. Risk Res.,* vol. 15, no. 8, pp. 991-1011, 2012. [http://dx.doi.org/10.1080/13669877.2012.686053]

[16] A. Edet, "An aquifer vulnerability assessment of the Benin Formation aquifer, Calabar, southeastern Nigeria, using DRASTIC and GIS approach", *Environ. Earth Sci.,* vol. 71, pp. 1747-1765, 2014. [http://dx.doi.org/10.1007/s12665-013-2581-y]

[17] S.M. Shirazi, H.M. Imran, S. Akib, Z. Yusop, and Z.B. Harun, "Groundwater vulnerability assessment in the Melaka State of Malaysia using DRASTIC and GIS techniques", *Environ. Earth Sci.,* vol. 70, pp. 2293-2304, 2013. [http://dx.doi.org/10.1007/s12665-013-2360-9]

[18] L.H. Yin, E.Y. Zhang, and X.Y. Wang, "A GIS-based DRASTIC model for assessing groundwater vulnerability in the Ordos Plateau, China", *Environ. Earth Sci.,* vol. 69, pp. 171-185, 2013. [http://dx.doi.org/10.1007/s12665-012-1945-z]

[19] S. Saidi, S. Bouri, and H. Ben Dhia, "Groundwater vulnerability and risk mapping of the Hajeb-jelma aquifer (Central Tunisia) using a GIS-based DRASTIC model", *Environ. Earth Sci.,* vol. 59, pp. 1579-1588, 2010. [http://dx.doi.org/10.1007/s12665-009-0143-0]

[20] A. Neshat, B. Pradhan, S. Pirasteh, and H.Z. Shafri, "Estimating groundwater vulnerability to pollution using a modified DRASTIC model in the Kerman agricultural area, Iran", *Environ. Earth Sci.,* vol. 71, pp. 3119-3131, 2014. [http://dx.doi.org/10.1007/s12665-013-2690-7]

[21] Z.A. Mimi, N. Mahmoud, and M. Abu Madi, "Modified DRASTIC assessment for intrinsic vulnerability mapping of karst aquifers: a case study", *Environ. Earth Sci.,* vol. 66, pp. 447-456, 2012. [http://dx.doi.org/10.1007/s12665-011-1252-0]

[22] B. Attoui, N. Kherici, and H. Kherici-Bousnoubra, "Use of a new method for determining the vulnerability and risk of pollution of major groundwater reservoirs in the region of Annaba-Bouteldja (NE Algeria)", *Environ. Earth Sci.,* vol. 72, pp. 891-903, 2014. [http://dx.doi.org/10.1007/s12665-013-3012-9]

[23] R. Li, and J.W. Merchant, "Modeling vulnerability of groundwater to pollution under future scenarios of climate change and biofuels-related land use change: a case study in North Dakota, USA", *Sci. Total Environ.,* vol. 447, pp. 32-45, 2013. [http://dx.doi.org/10.1016/j.scitotenv.2013.01.011] [PMID: 23376514]

[24] E. Sener, and A. Davraz, "Assessment of groundwater vulnerability based on a modified DRASTIC model, GIS and an analytic hierarchy process (AHP) method: the case of Egirdir Lake basin (Isparta, Turkey)", *Hydrogeol. J.,* vol. 21, pp. 701-714, 2013. [http://dx.doi.org/10.1007/s10040-012-0947-y]

[25] M.A. Mota Pais, I.M. Antunes, and M.T. Albuquerque, "Vulnerability mapping in a thermal zone, Portugal - a study based on DRASTIC index and GIS, Multi-disciplinary Research on Geographical Information in Europe and Beyond", *Proceedings of the AGILE'2012 International Conference on Geographic Information Science,* pp. 345-347 Avignon

[26] K. Tilahun, and B.J. Merkel, "Assessment of groundwater vulnerability to pollution in Dire Dawa, Ethiopia using DRASTIC", *Environ. Earth Sci.,* vol. 59, pp. 1485-1496, 2010. [http://dx.doi.org/10.1007/s12665-009-0134-1]

[27] C. Cox, and C. Madramootoo, "Application of geographic information systems in watershed management planning in St. Lucia", *Comput. Electron. Agric.,* vol. 20, pp. 229-250, 1998. [http://dx.doi.org/10.1016/S0168-1699(98)00021-0]

[28] C. He, "Integration of geographic information systems and simulation model for watershed

management", *Environ. Model. Softw.,* vol. 18, pp. 809-813, 2001.
[http://dx.doi.org/10.1016/S1364-8152(03)00080-X]

[29]   A.A. Millward, and J.E. Mersey, "Conservation strategies for effective land management of protected areas using an erosion prediction information system (EPIS)", *J. Environ. Manage.,* vol. 61, no. 4, pp. 329-343, 2001.
[http://dx.doi.org/10.1006/jema.2000.0415] [PMID: 11383105]

[30]   O. Nekhay, M. Arriaza, and L. Boerboom, "Evaluation of soil erosion risk using Analytic Network Process and GIS: a case study from Spanish mountain olive plantations", *J. Environ. Manage.,* vol. 90, no. 10, pp. 3091-3104, 2009.
[http://dx.doi.org/10.1016/j.jenvman.2009.04.022] [PMID: 19501954]

[31]   M. Trabucchi, C. Puente, F.A. Comin, G. Olague, and S.V. Smith, "Mapping erosion risk at the basin scale in a Mediterranean environment with opencast coal mines to target restoration actions", *Reg. Environ. Change,* vol. 12, pp. 675-687, 2012.
[http://dx.doi.org/10.1007/s10113-012-0278-5]

[32]   M. López-Vicente, A. Navas, L. Gaspar, and J. Machín, "Advanced modelling of runoff and soil redistribution for agricultural systems: the SERT model", *Agric. Water Manage.,* vol. 125, pp. 1-12, 2013.
[http://dx.doi.org/10.1016/j.agwat.2013.04.002]

[33]   Django project, *Django the web framework for perfectionists with deadlines.* https://www.djangoproject.com/

[34]   B. Borevsky, L. Yazvin, and L. Margat, Importance of groundwater for water supply.*Groundwater resources of the world and their use. IHP-VI, Series on Groundwater, N°6 Paris.,* I.S. Zektser, L.G. Everett, Eds., UNESCO, 2004, pp. 20-24.

[35]   A.C. Job, Ed., *Groundwater Economics.* CRC Press, Taylor & Francis Group: London, 2010, pp. 1-650.

[36]   SSD Foster, "Fundamental concepts in aquifer vulnerability, pollution risk and protection strategy", *Proceedings and information/TNO Committee on Hydrological Research,* vol. 38, 1987pp. 69-86

[37]   National Research Council, *Ground Water Vulnerability Assessment: Predicting Relative Contamination Potential Under Conditions of Uncertainty.* The National Academies Press: Washington, DC, 1993, pp. 1-204.

[38]   L Aller, JH Lehr, R Petty, and T Bennet, "DRASTIC: a standardized system to evaluate groundwater pollution potential using hydrogeologic settings", In: *National Water Well Association Worthington, Ohio, US EPA Report No.: 600/287/035, U.S. Environmental Protection Agency,* 1987.

[39]   J. Wang, J. He, and H. Chen, "Assessment of groundwater contamination risk using hazard quantification, a modified DRASTIC model and groundwater value, Beijing Plain, China", *Sci. Total Environ.,* vol. 432, pp. 216-226, 2012.
[http://dx.doi.org/10.1016/j.scitotenv.2012.06.005] [PMID: 22750168]

[40]   E. Fijani, A.A. Nadiri, A.A. Moghaddam, F.T. Tsai, and B. Dixon, "Optimization of DRASTIC method by supervised committee machine artificial intelligence to assess groundwater vulnerability for Maragheh–Bonab plain aquifer, Iran", *J. Hydrol. (Amst.),* vol. 503, pp. 89-100, 2013.
[http://dx.doi.org/10.1016/j.jhydrol.2013.08.038]

[41]   L. Xu, X. Xu, and X. Meng, "Risk assessment of soil erosion in different rainfall scenarios by RUSLE model coupled with Information Diffusion Model: A case study of Bohai Rim, China", *Catena,* vol. 100, pp. 74-82, 2012.
[http://dx.doi.org/10.1016/j.catena.2012.08.012]

[42]   D.D. Alexakis, D.G. Hadjimitsis, and A. Agapiou, "Integrated use of remote sensing, GIS and precipitation data for the assessment of soil erosion rate in the catchment area of "Yialias" in Ciprus", *Atmos. Res.,* vol. 131, pp. 108-124, 2013.

[http://dx.doi.org/10.1016/j.atmosres.2013.02.013]

[43]   G. Wordofa, *"Soil erosion modeling using GIS and RUSLE on the Eurajoki watershed Finland"*, Bacherol's dissertation. Tampere University of Applied Sciences, Department of Environmental Engineering, 2011.

[44]   R. Lar, Ed., *Soil Erosion Research Method.* Ankeny, Soil and water conservation Society, 1994.

[45]   M.A. Nearing, V. Jetten, and C. Baffaut, "Modeling responde of soil erosion and runoff to changes in precipitation and cover", *Catena,* vol. 61, pp. 131-154, 2005.
[http://dx.doi.org/10.1016/j.catena.2005.03.007]

[46]   WH Wischmeier, and DD Smith, *Predicting Rainfall Erosion Losses: A Guide to Conservation Planning with Universal Soil Loss Equation (USLE). Agriculture Handbook No. 537* US Department of Agriculture: Washington, DC, 1978, pp. 1-67.

[47]   K.G. Renard, G.R. Foster, G.A. Weesies, D.K. McCool, and D.C. Yoder, *Predicting Soil Erosion by Water: A Guide to Conservation Planning with the Revised Universal Soil Loss Equation (RUSLE). Agriculture Handbook.* US Department of Agriculture: Washington, DC, 1997.

[48]   R.P. Morgan, J.N. Quinton, and R.E. Smith, "The European Soil Erosion Model (EUROSEM): a dynamic approach for predicting sediment transport from fields and small catchments", *Earth Surf. Process. Landf.,* vol. 23, pp. 527-544, 1998.
[http://dx.doi.org/10.1002/(SICI)1096-9837(199806)23:6<527::AID-ESP868>3.0.CO;2-5]

[49]   D.C. Flanagan, M.A. Nearing, Ed., *USDA-water erosion prediction project: hillslope and watershed model documentation. NSERL Report No.: 10.* USDA-ARS National Soil Erosion Research Laboratory: Indiana, 1995.

[50]   E.G. Charles, C. Behroozi, J. Schooley, and J.L. Hoffman, "A method for evaluating ground-water-recharge areas in New Jersey", In: *New Jersey Geological Survey, Geological Survey Report GSR-32, Trenton,* 1993.

[51]   E. Custódio, and M.R. Llamas, *Hidrologia subterránea (Groundwater hydrology).* Omega: Barcelona, 1996.

[52]   W.H. Wischmeier, C.B. Johnson, and B.V. Cross, "A soil erodibility nomograph for farmland and construction sites", *J. Soil Water Conserv.,* vol. 26, pp. 189-193, 1971.

[53]   N.S. Loureiro, and M.A. Coutinho, "A new procedure to estimate the RUSLE EI30 index, based on monthly rainfall data and applied to the Algarve region, Portugal", *J. Hydrol. (Amst.),* vol. 250, pp. 12-18, 2001.
[http://dx.doi.org/10.1016/S0022-1694(01)00387-0]

[54]   G.F. Bonham-Carter, "Geographic Information Systems for Geoscientists, Modelling with GIS", *Computer Method in Geosciences,* vol. 13, pp. 152-153, 1994.

[55]   M.T. Pimenta, *Directrizes para a aplicação da equação universal de perda dos solos em SIG, Factor de Cultura C e Factor de Erodibilidade do Solo K. INAG/DSRH (Sistema Nacional de Informação dos Recursos Hídricos).*Lisbon, Portugal, 1998.

[56]   Z. Jiang, S. Su, C. Jing, S. Lin, X. Fei, and J. Wu, "Spatiotemporal dynamics of soil erosion risk for Anji County, China", *Stochastic Environ. Res. Risk Assess.,* vol. 26, pp. 751-763, 2012.
[http://dx.doi.org/10.1007/s00477-012-0590-0]

[57]   M. Fagnano, D. Nazzareno, I. Alberico, and N. Fiorentino, "An overview of soil erosion modeling compatible with RUSLE approach", *Rendiconti Lincei,* vol. 23, pp. 69-80, 2012.
[http://dx.doi.org/10.1007/s12210-011-0159-8]

[58]   S. Kumar, and S.P. Kushwaha, "Modelling soil erosion risk based on RUSLE-3D using GIS in a Shivalik sub-watershed", *J. Earth Syst. Sci.,* vol. 122, no. 2, pp. 389-398, 2013.
[http://dx.doi.org/10.1007/s12040-013-0276-0]

[59]   P.M. Carreira, J.M. Marques, and J. Espinha Marques, "Defining the dynamics of groundwater in

Serra da Estrela Mountain area, central Portugal: an isotopic and hydrogeochemical approach", *Hydrogeol. J.,* vol. 19, pp. 117-131, 2011.
[http://dx.doi.org/10.1007/s10040-010-0675-0]

[60]   J. Espinha Marques, J. Samper, and B. Pisani, "Evaluation of water resources in a high-mountain basin in Serra da Estrela, Central Portugal, using a semi-distributed hydrological model", *Environ. Earth Sci.,* vol. 62, pp. 1219-1234, 2011.
[http://dx.doi.org/10.1007/s12665-010-0610-7]

[61]   J. Espinha Marques, J.M. Marques, and H.I. Chaminé, "Conceptualizing a mountain hydrogeologic system by using an integrated groundwater assessment (Serra da Estrela, Central Portugal): a review", *Geosci. J.,* vol. 17, no. 3, pp. 371-386, 2013.
[http://dx.doi.org/10.1007/s12303-013-0019-x]

[62]   N. Ferreira, and G. Vieira, *Guia geológico e geomorfológico do Parque Natural da Serra da Estrela: Locais de interesse geológico e geomorfológico. Parque Natural da Serra da Estrela [Internet]* Ebook Library: Lisbon: Instituto da Conservação da Natureza, 1999, p. 12.

[63]   J. Espinha Marques, J.M. Duarte, and A.T. Constantino, Vadose zone characterisation of a hydrogeologic system in a mountain region: Serra da Estrela case study (Central Portugal).*Aquifers Systems Management: Darcy's legacy in a World of Impending Water Shortage.,* L. Chery, G. Marsily, Eds., Taylor & Francis: London, 2007, pp. 207-221.

CHAPTER 5

The Role of GIS and LIDAR as Tools for Sustainable Forest Management

Rubén Fernández de Villarán San Juan[*] and **Juan Manuel Domingo-Santos**

Departamento de Ciencias Agroforestales, University of Huelva, E-21819 Palos de la Frontera, Spain

Abstract: Regarding activities related to sustainable forests management, the spatial location of information is a very important factor, which requires tools capable of acquiring this data and handling them in a georeferenced format. For this reason, forest management has rapidly incorporated geospatial tools offered by new information technologies. Two important technologies used are Geographical Information Systems (GIS) and the remote sensing technology known as LIDAR (Light Detection and Ranging).Forestry applications of these technologies can be grouped into two broad categories: (i) Inventory and monitoring of natural resources; and (ii) Analysis and modeling of resources to facilitate sustainable planning and management. The first category is designed to measure the surface area, quantity, composition and condition of forest and natural resources of a management area. Thus, foresters use the LIDAR technology for acquiring digital information on the structure of the forest and the terrain; this information, properly processed with a GIS, helps analysts in assessing the health of the forest, calculating and classifying forest biomass, classifying land, or identifying soil drainage patterns, among other things. In the second category, once the above mentioned information has been mapped in a GIS environment, it is accessible to managers and researchers who can analyze and create models that optimize the decision-making on the resources under management, facilitating and optimizing forest planning. Therefore, wood felling can be scheduled in a sustainable way, as well as the design of firefighting infrastructures or the optimization of any other decisions related to use of resources or the protection of wildlife. This chapter aims to make the reader familiar with some variables of sustainable forest management, and with their integration into a GIS environment, as well as to introduce the basics of LIDAR technology and its powerful capabilities to acquire useful information for forest managers and planners.

Keywords: Digital Surface Models (DSM), Forest biomass appraisal, Forest management and planning, Geographical Information Systems (GIS), Geospatial tools, Light Detection and Ranging (LIDAR), LAS file.

[*] **Corresponding author Rubén Fernández de Villarán San Juan:** Departamento de Ciencias Agroforestales, University of Huelva, E-21819 Palos de la Frontera, Spain; Tel: +34959217620; E-mail: ruben@uhu.es

Ana Cláudia Teodoro (Ed.)

INTRODUCTION

Worldwide, forests are a source of products, water, biodiversity, ecological balance and human health plus life quality. Forest conservation is a major issue of environmental world policies; woodlands need protection against overexploitation, exposure to pollution, fire hazards, and deforestation intended for other land uses.

A milestone with regard to forest conservation and protection was the Rio de Janeiro United Nations Conference on Environment and Development (UNCED) celebrated in the year 1992, also known as the Earth Summit. The importance of forests for the environment and for a balanced economic development of humanity was there set up, thus concluding that the sustainable use of forests should be considered as a basic element of Nations' sustainable developments [1].

In 1993 at the second Ministerial Conference on the Protection of Forests in Europe (MCPFE) held in Helsinki, Resolution H1 on the "General Guidelines for the Sustainable Management of Forests in Europe" provided the following definition regarding a Sustainable Forest Management (SFM):

> [...], "sustainable management" means the stewardship and use of forests and forest lands in a way, and at a rate, that maintains their biodiversity, productivity, regeneration capacity, vitality and their potential to fulfill, now and in the future, relevant ecological, economic and social functions, at local, national, and global levels, and that does not cause damage to other ecosystems [2].

Addressing forest problems throughout the World has produced several forest management certification schemes, at regional as well as global levels; all of these schemes are based on certain principles or criteria which have been developed through indicators. An example of such a criterion set adopted by some certification schemes, is the pan-European criteria published for SFM at the Third MCPFE (Lisbon 1998), adopted following the technical meetings held in Geneva (1994) and Antalya (1995) [2]:

1. Maintenance and appropriate enhancement of forest resources and their contribution to global carbon cycles.
2. Maintenance of forest ecosystems' health and vitality.
3. Maintenance and encouragement of productive functions of forests (wood and non-wood).

4. Maintenance, conservation and appropriate enhancement of biological diversity in forest ecosystems.
5. Maintenance, conservation and appropriate enhancement of protective functions in forest management (notably soil and water).
6. Maintenance of other socio-economic functions and conditions.

In forestry properties, the basic tool to achieve healthy and useful forests, thereby to comply with all of the SFM criteria and indicators, encompasses the Management Plan (MP).

Management Plans

The MP is a document or a set of documents that should contain useful forest information for decision making and development of an action plan (see *e.g.* [3].).

Some MP items as directly related to the topic of this chapter will further be described briefly.

General Information Regarding the Managed Area

- Spotting of protected habitats and protected species: they need to be localized on a map, as well as the buffer areas that may suffer exploitation restrictions.
- Landform analysis: steepness, microclimatic conditions, view-points, view-sheds and other factors related to the relief set up of management constraints.
- Land cover: forests are divided into management areas depending on vegetation type, tree species, site quality, tree and shrubs densities, *etc.*
- Infrastructures: roads, tracks, footpaths, firebreaks, watering points, fences, and any other infrastructure need to be located, whenever useful for forestry management practices.

Forestry Resources Information

- Existing resources such as wood, fuelwood, biomass, cork, pine cones, oak acorns, mushrooms, grazing grass and shrubs, herbs or any other actual or potential productions.
- Tree density: one of the main management variables; density and age define the clearings and regeneration fellings.
- Canopy cover and basal area: along with tree density these variables allow to evaluate optimal land productivity uses.
- Wood volumes and growth: calculated from tree height and diameter, allow to compute estimates on wood yield.
- Biomass volumes and growth: modeled from several tree variables, allow to obtain estimates on biomass yield.
- Age structure of the forest: within each management area and in the whole of the

forest managed, age structure will condition the silvicultural and management systems.

• Forest fire risk assessment, depends on terrain and meteorological variables as well as on vegetation and stocked fuel conditions.

Action Plans for SFM

• Calculating the sustainable yield of wood, biomass or any other forest products.
• Localizing and quantifying the log fellings planned on a yearly basis during 5 up to 10 years. All restrictions regarding biodiversity, landscape, soil erosion or any other limitation will be integrated into this planning, in order to determine felling areas as well as felling techniques.
• Infrastructure development for fire protection and recreational use.
• Infrastructure development for resource exploitation.
• Localizing and quantifying any other actions for forest improvement, wildlife protection, *etc*.

TOOLS FOR INVENTORY AND MANAGEMENT PLANS

As has already been noted, resources information entails a basic step to manage resources in a sustainable way. Information is obtained by means of forest inventories, based on forest sampling plot measurements together with statistical analysis of the sampled data.

In order to measure sampling plots, a team of 2 to 4 persons must localize the plot's center and identify the trees inside of the plot, measure the trees and count them. Staff may also mark down any other relevant information on the ecological features of the sampled area.

Plot measurements are time consuming and costly, so the number of measured plots must be limited to that strictly necessary. Depending on the variability of the forest characteristics, the number of plots may vary from about one plot per hectare to one plot every 25 hectares.

Ever since invented, remote sensing has broadly been applied in order to reduce plot samplings. Satellite or airborne multispectral images (MS images) are very useful to identify tree species, canopy cover, vegetation health, terrain elevation and other useful data (see *e.g.* [4, 5].). Nevertheless, these images have low performance in providing most of the quantitative information concerning forests (*e.g.* [6].), as listed before.

The LIDAR (Light Detection and Ranging) technology has been a revolutionary tool to obtain a huge amount of terrain data. It does not only obtain terrain elevations but in addition also gives the terrain surfaces covered by each vegetation layer (grass, shrubs and trees). Table **1** presents the referred management variables as well as the tools needed or useful to obtain them.

Table 1. Potential application of management tools to evaluate variables of forestry management plans.

Variable	On-site plots	MS Images	LIDAR	GIS Analysis
Protected areas		A		N
Landform analysis			A	N
Landcover	A	A	C	P
Infrastructures		A		N
Existing resources identification	N			
Tree density	N	A	B	P
Canopy cover-basal area	N	A	B	P
Wood volumes – growth	N		B	P
Biomass volumes and growth	N		B	P
Age structure	N		A	P
Fire risk assessment	N	C	A	P
Sustainable yield	N		A	P
Felling areas				N
Infrastructure planning				N

A: aid, very helpful; C: complementary aid; N: necessary; B: alternative to a needed tool; P: processed from LIDAR data or image processing.

As can be noted from Table **1** the LIDAR data should be matched to some reference plot data, in order to calibrate the system and so to minimize the errors. Anyway, the applicability of the LIDAR technology may drastically reduce the number of sampling plots, as long as the LIDAR data are available at an affordable cost.

This chapter aims to present the basics of LIDAR technology as well as its actual applications, combined with GIS, for the improvement of the quality and quantity of data used in forest management and planning. These improvements may also cause a significative cost reduction on data acquisition, which may lead to a better monitoring of all kind of forests.

THE LIDAR TECHNOLOGY

The LIDAR technology is an active detection technique which uses the same principle as the RADAR technology, yet instead of using radio waves uses a laser beam. While the RADAR uses radio waves to determine an object's distance by measuring the time delay between a pulse transmission and the detection of the reflected signal, the LIDAR makes use of much shorter wavelengths of the electromagnetic spectrum.

The wavelength used by the laser of the LIDAR systems may vary between near infrared (NIR), ultraviolet (UV) and visible (VIS), according to the measurement's objective. While to measure short and medium distances (topographic works) a wavelength of 1064 nm (NIR) is utilized, for bathymetric measurements the wavelength must be adjusted to 532 nm (green) [7], as displayed in Fig. (**1**).

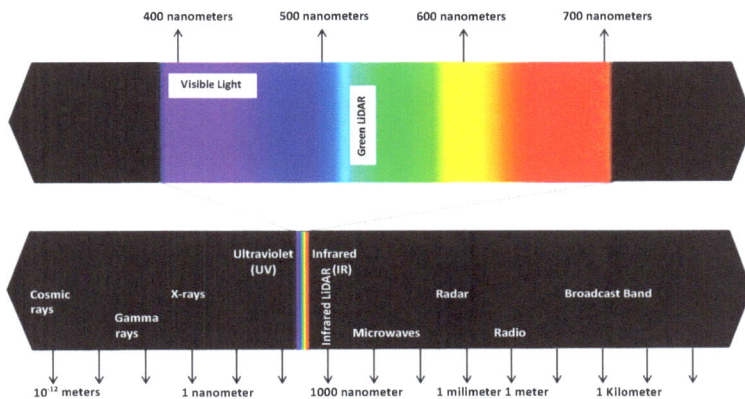

Fig. (1). Electromagnetic Spectrum indicating the most usual Wavelengths of the LIDAR Systems.

Therefore an object's distance is determined by measuring the time delay between emission of a light pulse and detection of the pulse reflected back from the object (Fig. **2**), taking into account that the laser light pulse travels at the speed of light.

With the intention of mapping large areas, the possibility exists to mount the LIDAR system on an aerial platform (airplane, helicopter or drone), from which to send the light pulse to the Earth's surface; some of these pulses bounce back to the LIDAR sensor. The time taken by the laser beam to reach the Earth's surface and bounce back to the sensor in the air transport is used to estimate the Earth surface element's distance.

The LIDAR scanner is composed not only of the transmission and reception laser, in addition it also presents a differential satellite global positioning system (DGPS), which is applied to determine the platform's location, together with an inertial measurement unit (IMU) which enables to determine the platform's angle (pitch, roll, yaw) (Fig. **3**). The information provided by the three systems, once integrated mathematically, provides a point cloud which records for each item the horizontal coordinates (x,y) and the elevation (z).

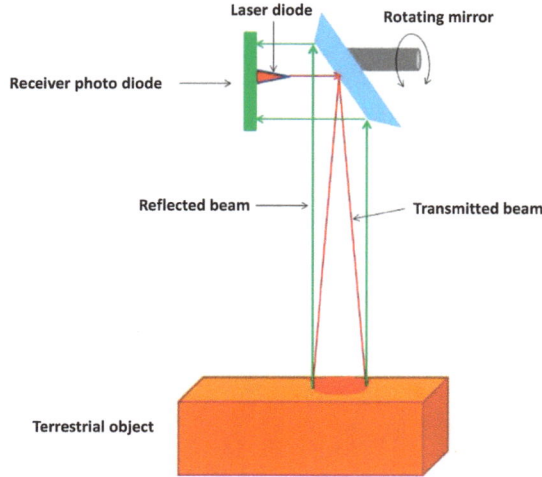

Fig. (2). Diagram of a LIDAR scanner and its working principle.

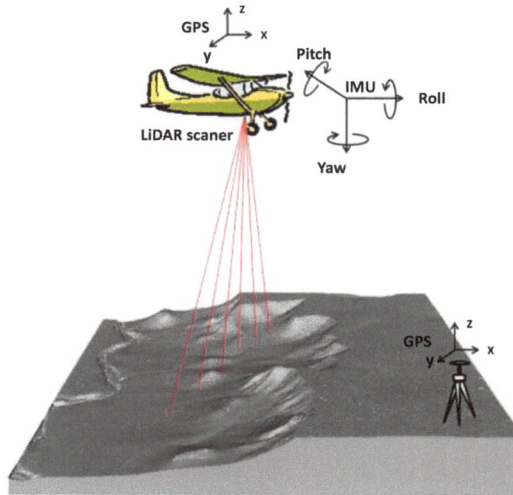

Fig. (3). Components of a LIDAR system.

This type of technology allows to directly measure three-dimensional structures and terrain surfaces. Depending on the method used to capture data, extremely dense point clouds can be obtained compared to other capture methods (up to five points per square meter). The result of opting for such a high resolution technology entails a higher measurement precision of the terrain elevations and the ground objects' heights. This feature is one of the foremost advantages of the LIDAR technology in comparison to other conventional optical instruments, such as digital cameras. On this respect one of the first things to draw the attention to this technology encompasses the effective altimetry accuracy which can be retrieved. Although the exact precision remains unknown, studies that have so far been conducted agree that the altimetry accuracy achieved is higher than that obtained with planimetry, given that for extended areas the mean square error of the planimetric coordinates [x,y] is estimated to amount to about 15 cm, while for the "z" coordinate errors below 8 cm can be attained [8]. This accuracy can also be achieved through photogrammetry, although this option is more expensive and requires much more time handling and processing data.

A laser pulse produced by a LIDAR system embodies a known finite diameter, therefore it could be possible that only part of the laser's diameter was reflected on a certain object, while the rest of the pulse would continue to travel until reflecting on another object, thus the original pulse would have produced a second reflection. Each pulse can in principle generate several reflections or returns as can be seen in Fig. (**4**).

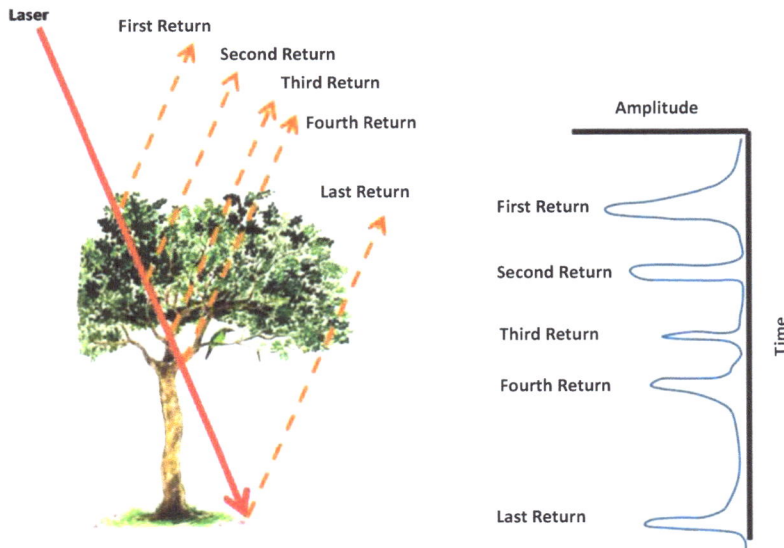

Fig. (4). Various reflections generated by a laser pulse emitted by the LIDAR System.

Initially, LIDAR systems were only capable of recognizing a single return, nonetheless currently they can measure several returns per pulse (from 3 to 5) [9]. These multiple returns can be used to study and to analyze the information about the objects on the Earth's surface. The signal of the first pulse allows describing the surface or upper part of an object, while the signal of the last eco is typically used to measure the soil surface.

Obviously with more returns per pulse registered the more accurate will be the information retrieved from the scanned surface, but the LIDAR dataset will tend to be very large, simply owing to the large number of pulses emitted by the laser scanner. Therefore most LIDAR systems have been designed to record up to five returns per pulse, given that in practice it has been established that the fourth and fifth pulse rarely occur.

The data quality and quantity obtained through a LIDAR flight depend directly on some parameters of the laser scanner and on the flight's configuration, given that these determine the point density captured by the LIDAR system. The flight configuration parameters with a higher influence on quality are depicted in Fig. (5) and listed below [10]:

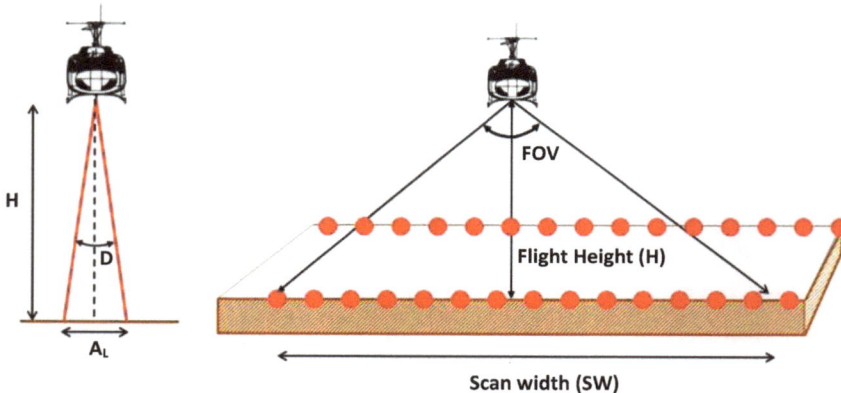

Fig. (5). Main flight parameters which sustain an influence on the LIDAR data quality obtained.

Pulse Frequency (F): Corresponds to the number of laser pulses that the sensor is capable of emitting per second, usually in the range between 200,000 to 400,000 pulses per second (200 to 400 KHz).

Scanning Frequency (fc): Number of scans per second or scanned lines per second, usually 25 to 90 Hz.

Field of View (FOV): Maximum angle at which the laser beam is emitted perpendicular to the flight direction. Generally the low field of view angles are more interesting, since usually with smaller FOV a higher product quality is obtained, since for example very large angles increase the footprint's size with respect to the size of the footprint in the nadir.

Flight Height (H): Corresponds to the scanning's executed height. Depends on the laser transmission's power and also on the mobile platform operated.

Maximum Scan Width (SW): Maximum scan width transversal to the flight direction, depends on the FOV and the flight height.

Divergence of the Laser Beam (D): Aperture angle of the laser beam.

Laser Footprint (Al): Illuminated Surface by a laser beam. Depends on the beam's divergence and the flight height.

Return Density per Surface Unit: Corresponds to the number of returns per surface unit.

Considering forest areas, the first pulse will impact on the top canopy allowing to characterize forest variables related to mass height or the canopy cover fraction. Part of the pulse can generate second, third or further rebounds while reflected on other plant parts such as leaves, branches or trunks, which allow to study or characterize forest variables, such as the canopy volume, the number of trees per hectare or other density variables. The last rebound registered matches usually the part of the laser beam reflected on bare soil, allowing to create topographic maps or terrain elevation models.

Recently, the LIDAR systems used to measure forest variables have evolved, allowing to be greatly improved [11, 12] in this manner they have incorporated laser beams of a greater diameter or footprint (diameters of 5m up to 25m) being able to receive and process the complete pulse wave, instead of only storing pulse returns. These improvements consent to attain a better characterization of the intermediate parts of the canopy and therefore of the forest's vertical structure (Fig. **6**).

LIDAR systems, in addition to recording the time lag between the pulse emission and reception, are also able to store information about each bounce's intensity or energy amount reflected, *i.e.* the laser beam's return force after being reflected by an object. Consequently, high reflectivity objects as could be snow or metal roofs evidence high intensity or energy returns, while objects of a low reflectivity as for instance roads return low energy or low signal strength pulses. Water bodies absorb most of the laser beam energy and no rebounds are generated, except if adequate wavelengths are radiated (near to the green region of the spectrum).

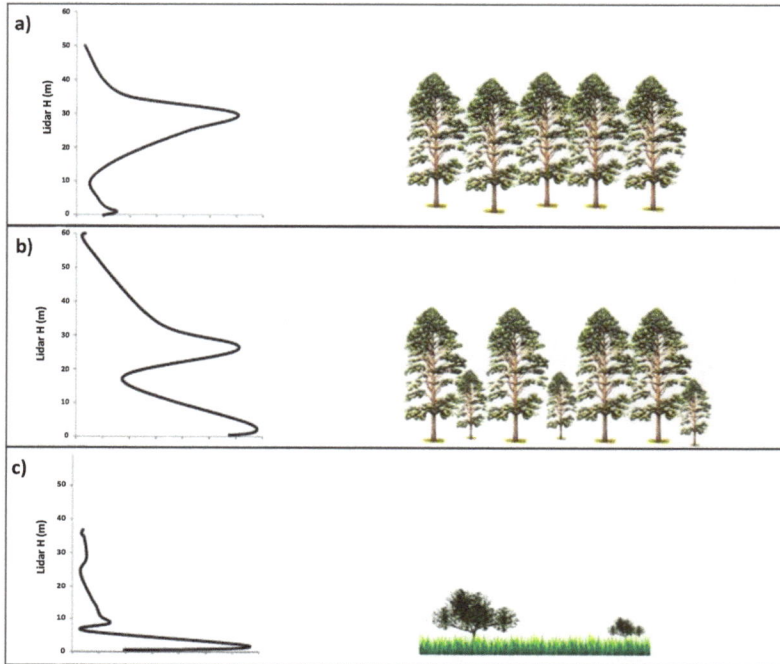

Fig. (6). Laser beam response to different mass structures: (**a**) a regular mass of a single layer (**b**) a mass with two height layers (**c**) absence of trees, bare soil with a bit of scrubland.

This rebound intensity information can be used to generate an image; usually an 8 bit color depth (256 colors) which applied to a grayscale palette allows to visualize point clouds similar to those of a panchromatic orthophotography (Fig. 7). This orthophotograph once properly treated with a geographic information system, allows to generate a raster layer or orthophotograph of a certain precision, each pixel's digital value containing the average of all the signal rebounded intensities falling within a same pixel.

However, it should be noted that the digital value stored within a pixel corresponds to the object's reflectivity according to the laser wavelength applied being the derived intensity data set not standardized, therefore its use is limited to create orthophotographic images which prevent to retrieve any other information. An interesting advantage given that the LIDAR system uses an active sensor, image intensities do not depend on the prevailing light conditions, allowing to capture data under cloudy conditions and even nocturnally.

Fig. (7). Rebound intensity image of the LIDAR system.

The information gathered by the LIDAR sensor is stored in a file type "Laser file format Exchange activities" also known as LAS file, whose specifications have been developed by the American "Society for Photogrammetry & Remote Sensing". The LAS file is encoded in a public exchange format comprising three-dimensional point type data, each point storing as minimum: the x and y coordinates together with the registered height, the rebound intensity, the emitted pulse number, the reflected pulse number, last, the pulse emission angle.

INFORMATION PROVIDED BY LIDAR

The LIDAR technology presents many advantages compared to the other traditional cartographic techniques, such as photogrammetry. It possesses the ability to simultaneously record information on the structure of the vegetation cover and on the soil surface that exists underneath of the tree canopy, an unthinkable question attained by using the traditional techniques of photo-interpretation. This sampling accuracy increase of the soil surface substantially allows to improve the ground topographic analysis achieving a better understanding of the geographic features.

In this manner the first product that can be obtained by the LIDAR information consists of a digital terrain model (DTM), which simply can be defined as a numeric data structure representing the spatial distribution of the variable ground height. Generation is achieved using the rebounds reflected on the bare ground, regardless of whether they are the first, second or third rebound.

The next information derived from the LIDAR point data clouds are the digital surface models (DSM) which correspond to digital terrain models that include the elevations of existing objects on the Earth's surface, such as trees, buildings or the bare soil with no previous coverage [13]. This surface model is of particular interest to the forestry sector, as it offers a clear view of the highest canopy surface of wooded areas. This model is generated, logically from the first rebound of all the pulses [14], (Fig. **8**).

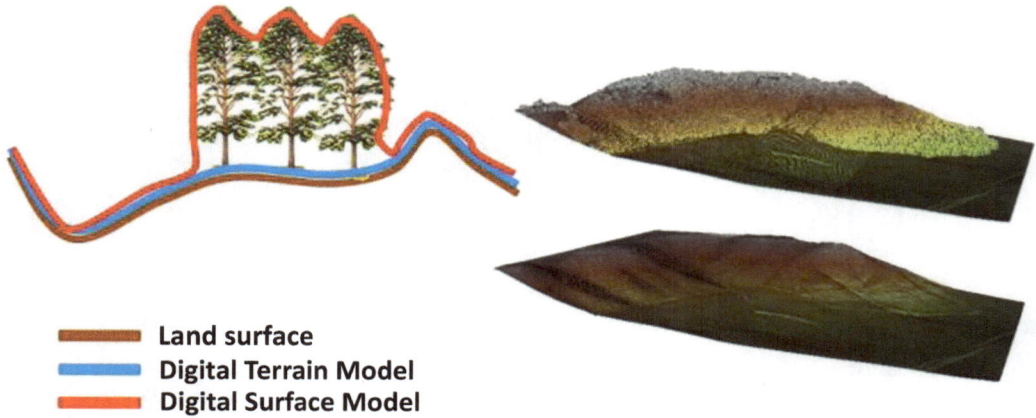

Land surface
Digital Terrain Model
Digital Surface Model

Fig. (8). Digital Terrain (DTM) and Surface (DSM) Models: DSM (red line and upper 3-D image) includes canopy cover, whereas DTM (blue line and lower 3-D image) represents the bare ground landforms.

In contrast to the two previous models a third product of great forestry interest is obtained consisting of the vegetation's digital model containing the canopy heights read by the LIDAR scanner (Fig. **9**).

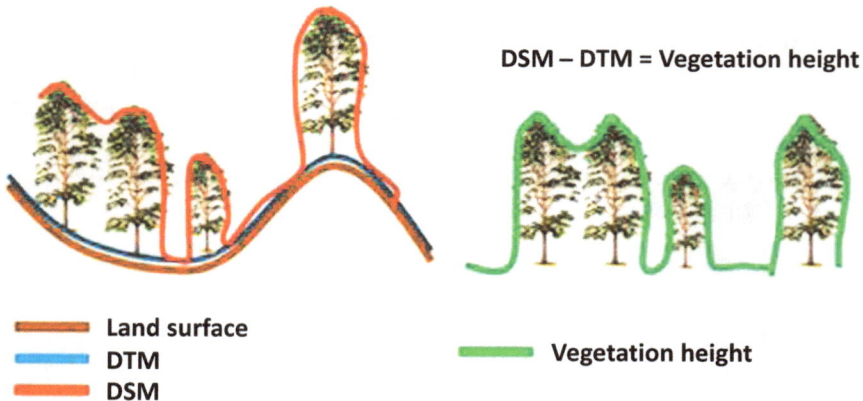

DSM – DTM = Vegetation height

Land surface
DTM
DSM

Vegetation height

Fig. (9). Standardization of the Vegetation height recorded by the LIDAR scanner.

Handling LIDAR data applied to forest inventories can potentially provide forest variables at an individual tree level, allowing to obtain a broader picture of the forest's real structure. These forest mass variables usually are: height, canopy fraction and tree profiles. These attributes can be used to derive other forestry measures such as the bi-symmetrical area and wood volume, as well as the biomass amount for energy production and the analysis of forestland carbon capture.

THE FORESTRY INVENTORY WITH LIDAR

Undoubtedly forestry is one of the fields in which the LIDAR technology has exhibited its highest potential, thanks to this system's ability to generate dense three-dimensional point clouds. LIDAR data provides horizontal and vertical information with a high spatial resolution and accuracy, with forest attributes impossible to be obtained by any other remote means [15, 16]. Numerous authors have presented results demonstrating the LIDAR system's capability of measuring dasometric forest mass variables and individual trees dendrometric measurements [17 - 24].

Giving the main points on the forest variables obtainable or directly derived from the LIDAR data would be: canopy height, underlying canopy topography, aboveground biomass, basimetric area, average trunk diameter, treetop volume or vertical distribution of the vegetation [25 - 34].

Prior to obtaining any forest variables resulting from a LIDAR flight, a pre-processing of LAS files is always needed, which usually consists of a point cloud filtering (identification of bare soil returns and identification of returns produced by objects on the ground), the MDE generation and the standardization of each return's height with respect to the ground (or creation of the MDV), followed by a statistical analysis of the rebounds above a determined height (metrics of the LAS file) (Table **2**).

Table 2. Statistical Variables with forestry interest derived from a LAS file (flight metrics).

Statistical Variables derived from the laser return above a certain height	Forestry Utility
Percentage of first returns above X meters	Canopy cover (%)
Minimum Elevation	Lowest canopy height
Maximum Elevation	Highest canopy height
Mean Elevation	Average Canopy Height
Elevation mode	Canopy Height Mode
Elevation Standard deviation	Forest regression models

(Table 2) contd.....

Statistical Variables derived from the laser return above a certain height	Forestry Utility
Elevation Variance	Forest regression models
Elevation Coefficient of variation	Forest regression models
Elevation Coefficient of skewness	Forest regression models
Elevation Interquartile range	Forest regression models
Elevation Coefficient of Kurtosis	Forest regression models
Elevation Average Absolute Deviation	Forest regression models
Elevation Percentile 1 to 99	Forest regression models

A significant number of the main forestry attributes can be directly obtained from the (X, Y, Z) data provided by the LIDAR system LAS files; namely some of these attributes could be: Individual tree height, Average canopy height, Dominant height, Canopy cover, Number of trees, Crown size, Height to live crown, Crown volume.

Other variables must be obtained indirectly by establishing statistical models through empirical relationships, namely *e.g.* Wood volume, Biomass, CO_2 stocked in biomass, Quadratic mean diameter, Basal area, Leaf area index, Crown density.

As already mentioned above the forest variable estimates derived from LIDAR data can be processed resorting to either of two approaches, depending on the rebound densities obtained in LIDAR flights.

The first approach involves evaluating forestry variables at the individual tree level which necessitates of flights capable of obtaining high point or rebound densities comprising 5 to 10 rebounds per m2 [35]. The second approach corresponds to estimating forestry variables for an entire area or woodland, be it a piece of inventory, a stand or a complete scrubland; corresponding usually to flights covering low point densities containing 0.5 to 4 rebounds per m^2 [36, 38].

The most pronounced difference between these two approaches encompasses that the first method or tree method is based on the detection and delineation of individual trees to subsequently apply allometric equations at the individual level; in contrast the second method or mass method uses directly the returns reported by the LIDAR sensor onto the work surface (plot, stand or scrubland) in order to establish statistical relationships allowing to estimate the forest variables of interest.

The mass method attempts to reflect both the vertical and the horizontal space organization of the forestry mass components. This type of inventory is possible owing to the LIDAR point cloud describing in detail and in a continuous manner the vegetation structure as can be appreciated in Fig. (**10**).

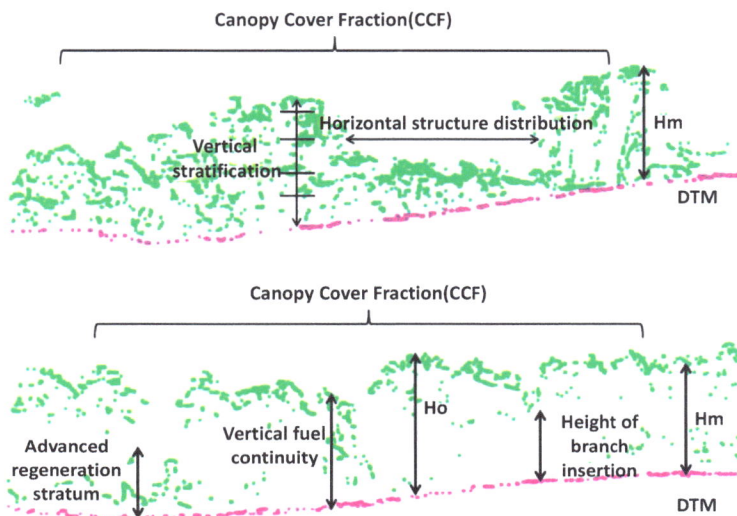

Fig. (10). Possible description of the horizontal and vertical mass structure.

In this inventory type the objective is based on knowing along the entire scrubland surface the following forestry variables: basimetric area, dominant height and wood volume among other things.

In general the LIDAR inventories carried out by the mass method are considered as double sampling inventories. The objective of this inventory method consists in estimating determined forest variables, such as the basimetric area whose measurement results costly, by means of exploiting its relationship with other auxiliary variables easier and more economic to measure; in this case, the basimetric area can be estimated using percentiles, the standard deviation of the point cloud's height distribution or the rebound percentage above a certain height (Fig. **11** and **12**).

Numerous studies seem to have obtained good results with respect to various forestry variables, determining their value with an acceptable accuracy, for example, Means *et al*. [19] used the height distribution percentiles and the canopy cover fraction to estimate the mean mass height, wood volume and basimetric area of a Pseudotsuga douglasii forest, with tree heights ranging between 7 to 52 m, obtaining models with coefficients of determination (R^2) of 0.93 and 0.97 and 0.95 respectively for the average height, the timber volume and basimetric area.

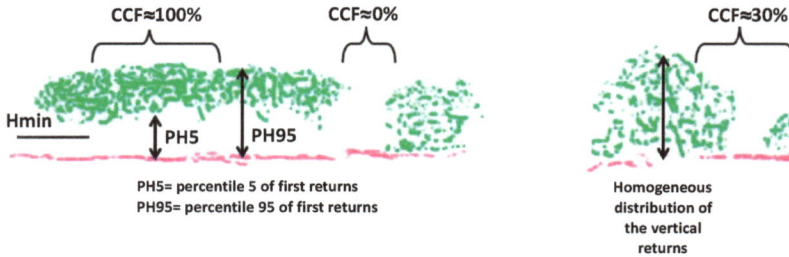

Fig. (11). LIDAR rebound distribution with respect to the vegetation and its relationship with canopy cover fraction (CCF).

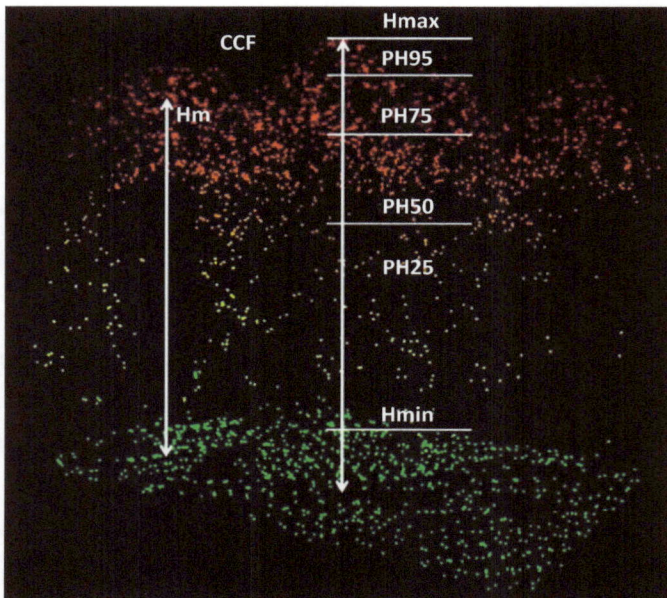

Fig. (12). Relationship between the first rebound percentiles with the canopy cover fraction (CCF) and the mass height.

In Spain, García *et al.* [39] applied this methodology to conduct a forest inventory, attaining equations of total volume, aboveground total biomass, basimetric area and trees per hectare with R^2 values of 0.90, 0.89, 0.89 and 0.80, respectively.

As has already been indicated in the introduction, traditional forest inventories determine the object forestry variables by sampling a small percentage of the surface (sampling plots) to subsequently extrapolate the results to the entire area, by means of other auxiliary variables. The LIDAR inventories require two sampling phases: a first phase carried out on ground which measures or estimates the variable to be modeled in a representative area sample and a second phase which measures the auxiliary variables or LIDAR data, to subsequently establish the correlations between the measurements and the LIDAR data through regression models (Fig. **13**).

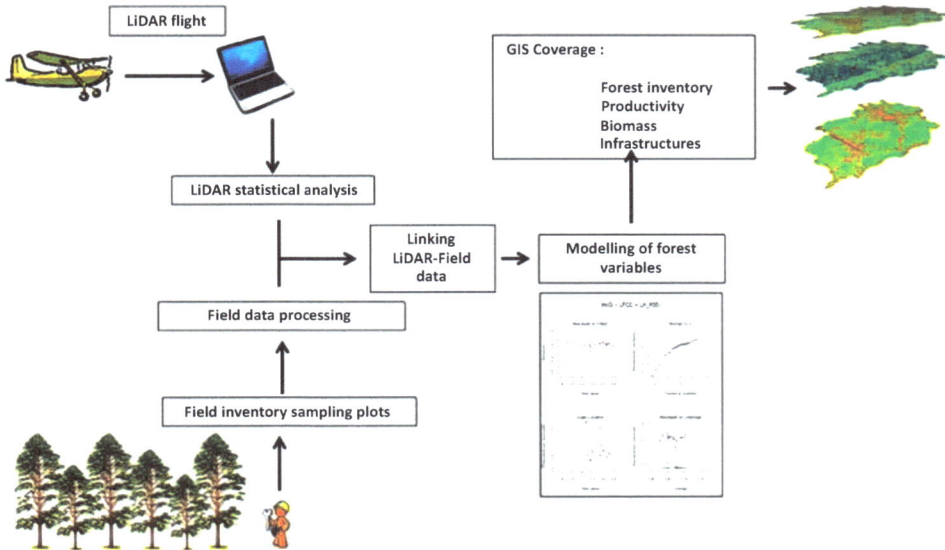

Fig. (13). General workflow of a forestry inventory using LIDAR.

In this manner a traditional forest inventory will have an associated error derived from the sampling fraction, whereas a LIDAR inventory will lack a sampling error as variables are measured throughout the entire mass surface, not being restricted to a fraction. However, the LIDAR inventory is not devoid of errors, since the data's extrapolation is associated to the goodness of fit of the regression models.

Thus, inventories by traditional plot samplings offer good estimates over large management units or strata yet bad estimates in smaller units (inventory unit or stand), on the other hand, LIDAR inventories allow to substantially improve the error estimation of forest variables in smaller management units, making it easier to improve the planning actions at a stand scale and to apply more competently flexible management methods that respond much better to the multifunctional demands that society is requiring out of woodlands.

The main differences between a LIDAR inventory and a classic inventory by sampling can be observed in Table **3**:

Table 3. Main differences between forestry inventories with LIDAR and a classical inventory by sampling.

Characteristic	LIDAR	By sampling
Information volume	Up to 40.000 vegetation height figures per ha	No more than 10 height and diameter figures per ha
Sampling Fraction	100%	Not more than 4%
Errors	Depend only on the quality of the regression models (plus possible instrumental errors).	Depend on the sampling intensity (number of plots) and on the quality of the regression models (plus possible instrumental errors).
Results	Provides details at a local scale and of the entire mass. Problems to distinguish species and diametric classes.	Provides robust data for complete management units, not at a local scale (stands or inventory units. Possibility to distinguish species and diametric classes.
Costs	High fixed cost (flight and processing), but covering extensive surfaces low cost per surface unit (possible multi-functionality of the LIDAR data for large areas).	Low fixed cost, mainly variable cost; with larger surfaces higher costs. In general, high-cost per surface unit.

THE LIDAR DATA SOURCES

This section wants to convey that in addition to the LIDAR data obtained from a planned flight and adjusted to the work's objectives, currently a series of public access Internet data sets are available, which may be useful with respect to some works. It must be borne in mind that the data set obtained by a planned flight tends to present a much higher quality (usually) compared to the free access data, given that the point densities obtained by the latter tend to be low and do not reach the minimum required for a forest inventory (over 0.8 points/m^2 respective to mass methods and 1.2 points/m^2 for individual trees). Nonetheless, these freely available data sets comprise a very good quality in order to obtain useful information suited for other aspects of forestry planning such as digital terrain models or forestry infrastructure maps.

The online access to the public LIDAR data varies significantly among the different spatial data infrastructures (SDIs); nevertheless most SDIs possess an interactive system which allows to select data from a particular area.

Within the European Union, notably owing to the community 2007/2/EC directive to establish an Infrastructure for Spatial Information in the European Community

(INSPIRE), member States have carried out LIDAR flights over their territories, aiming to obtain high precision terrain digital models. Availability of the raw flight data (LAS data) is uneven among the different countries, given that each country has applied their own criteria regarding quality and geoportal features.

The United States is the country which produces and facilitates most of the LIDAR data; several sites offer free access data, even though all the information is recollected in the "National LIDAR Dataset" project available through the public site of the United States Interagency Elevation Inventory (USIEI).

At a global level two very interesting initiatives can be cited which aim to create a social network to share LIDAR data: i) Open Topography, depending on the University of California; ii) online-LIDAR community managed by the company Dielmo.

When the public data does not adapt to the different objectives set out by the work to be carried out, a specialized company running a LIDAR flight specific to the work area of interest should be hired. Prior to its execution the planning of the flight should be undertaken, taking into account the various flight parameters which ensure that the data captured is sufficient in quantity and quality to generate the required models. The LIDAR flight parameters which mostly influence on the obtained data quality include: the pulse frequency, the scan frequency, the scanner's field of vision, the flight height, the flight speed and the laser's footprint, given that all these parameters affect the retrieved resolution or point density per surface.

THE SOFTWARE FOR LIDAR DATA TREATMENT AND INTEGRATION

It should be taken into account that the LIDAR data treatment and management applications are continuously being renewed, particularly when such a novel and innovative subject as the LIDAR systems are considered. Therefore, Table **4** intends to give a general outline of the most featured and commercially available software packages.

Table 4. Some examples of software for LIDAR data treatment.

SOFTWARE NAME	Creator	Processing	Source	Observations
TerraScan	Terra Solid	Whole	C	
BCAL LIDAR Tools	Boise State University	Standard	O	Complement to the remote sensing ENVI software package

(Table 4) contd.....

SOFTWARE NAME	Creator	Processing	Source	Observations
FugroViewer	Fugro	Standard	O	Displays other vector/raster data
Fusion/LDV	USDA	Standard	O	High quality for DTM and DSM
LASTOOLS	Rapidlasso GmbH	Standard	C	
ArcGis	ESRI	Standard	C	
Global Mapper – LIDAR Module	Blue Marble Geographics	Standard	C	

Whole processing: all processing, including calibration of the LIDAR sensors and data quality control; Standard processing: classify, convert, filter and transform the point clouds to a raster image; C: Commercial; O: Open source, free software.

CONCLUSIONS

Decision taking processes in forest management and planning need information on a wide range of tree and vegetation variables, infrastructure, landforms, *etc*. All this information is nowadays managed by means of GIS.

Forest inventory variables can be obtained on-site by measuring forest variables in sampling plots. It is a costly and time consuming work; some forest areas may be very difficult to access. As forests grow and change, inventories need to be updated every 5 to 10 years.

Multispectral remote sensing images have been used on forest management and planning ever since they were available. These images are very helpful on identifying different land cover areas, locating infrastructures, *etc*. GIS platforms allow viewing and processing (manually or automatically these images). However accurate quantitative information on forest mensuration is not provided by remote sensing images.

LIDAR technology uses the reflection of a laser beam signal, broadcasted from an aircraft, to scan a 3-D model of the terrain, vegetation or any other objects covering the terrain surface.

As trees and vegetation in general are not opaque solids, the data obtained need to be classified, filtered and converted by means of algorithms implemented through the adequate software; most GIS platforms have tools developed for this purpose.

The processed LIDAR data provides directly forest variables such as Individual tree height, Average canopy height, Dominant height, Canopy cover, Number of trees, Crown size, Height to live crown, Crown volume. Other volumetric variables like wood or biomass volumes can be inferred from regression models based on data from on-site plot samplings.

LIDAR data must be sufficient in quantity and quality to generate the surface models; the flight parameters will prompt the utility of data for forest inventory.

LIDAR technology decreases dramatically the cost of forest inventories for broad areas of managed forests. It does not mean the disappearance of fieldwork but its reduction, especially for inventory updates, when a basic on-site sampling has been carried out previously.

High-performance LIDAR technology involves bulky, heavy components that usually require manned aerial platforms to operate. However, the development of Micro-Electro Mechanical Systems (MEMS) has made it possible to reduce and lighten LiDAR systems, facilitating their integration into micro unmanned aerial vehicles (UAVs or drones), where they can be combined with other remote sensing devices such as infrared or visible spectrum cameras. These systems have very low flight height, so they can provide greater precision than manned platforms, but their efficiency will be much lower because they have a narrower scan width. They can be very useful for providing detailed, low-cost information on the inspection of logging or any other forestry work at the local level.

LIDAR information implemented on a GIS platform can be considered a revolution in forest management and planning; this technology will allow proper sustainable management of natural areas or of low profit forests that cannot afford classical inventories. It will also provide a powerful tool for the conservation of natural protected areas threatened by deforestation or illegal practices, even when these areas are difficult to access.

CONSENT FOR PUBLICATION

Not applicable.

CONFLICT OF INTEREST

The author (editor) declares no conflict of interest, financial or otherwise.

ACKNOWLEDGEMENTS

Declared none.

REFERENCES

[1]   United Nations, "The World Conferences: Developing Priorities for the 21st Century", *UN Briefing Papers,* 1997

[2]　Forest Europe, *The Ministerial Conference on the Protection of Forests in Europe.* http://www.foresteurope.org/

[3]　P. Bettinger, K. Boston, J.P. Siry, and D.L. Grebner, *Forest Management and Planning.* Academic Press, Elsevier: Burlington, MA, USA; San Diego, CA, USA; London, UK, 2009.

[4]　E. Tomppo, H. Olsson, G. Ståhl, M. Nilsson, O. Hagner, and M. Katila, "Combining national forest inventory field plots and remote sensing data for forest databases", *Remote Sens. Environ.,* vol. 112, pp. 1982-1999, 2008.
[http://dx.doi.org/10.1016/j.rse.2007.03.032]

[5]　R.H. Wynne, Forest Mensuration with Remote Sensing: A Retrospective and a Vision for the Future, *Southern Forest Science: Past, Present, and Future. United States Department of Agriculture Forest Service.Southern Research Station. General Technical Report SRS-75.,* H.M. Rauscher, K. Johnsen, Eds., 2004.

[6]　J. Hyyppä, H. Hyyppä, M. Inkinena, M. Engdahla, S. Linkob, and Y.H. Zhuc, "Accuracy comparison of various remote sensing data sources in the retrieval of forest stand attributes", *For. Ecol. Manage.,* vol. 128, pp. 109-120, 2000.
[http://dx.doi.org/10.1016/S0378-1127(99)00278-9]

[7]　M.A. Lefsky, W.B. Cohen, D.J. Harding, G.G. Parker, S.A. Acker, and S.T. Gower, "Lidar remote sensing of aboveground biomass in three biomes", *Glob. Ecol. Biogeogr.,* vol. 11, no. 5, pp. 393-399, 2002.
[http://dx.doi.org/10.1046/j.1466-822x.2002.00303.x]

[8]　J. Estornell Cremades, *"Análisis de los factores que influyen en la precisión de un MDE y estimación de parámetros forestales en zonas arbustivas de montaña mediante datos LIDAR",* PhD dissertation. Universidad Politécnica de Valencia, Valencia, 2011.

[9]　J.S. Evans, A.T. Hudak, R. Faux, and A.M. Smith, "Discrete Return Lidar in Natural Resources: Recommendations for Project Planning, Data Processing, and Deliverables", *Remote Sens.,* vol. 1, pp. 776-794, 2009.
[http://dx.doi.org/10.3390/rs1040776]

[10]　D. Gatziolis, and H.E. Andersen, "A guide to LIDAR data acquisition and processing for the forests of the Pacific Northwest", In: *Gen. Tech. Rep. PNW-GTR-768* U.S. Department of Agriculture, Forest Service, Pacific Northwest Research Station: Portland, OR, 2008, p. 32.
[http://dx.doi.org/10.2737/PNW-GTR-768]

[11]　J.B. Blair, D.B. Coyle, J.L. Bufton, and D.J. Harding, "Optimization of an airborne laser altimeter for remote sensing of vegetation and tree canopies", *Proceedings, International Geoscience and Remote Sensing Symposium,* pp. 939-41 Pasadena CA
[http://dx.doi.org/10.1109/IGARSS.1994.399307]

[12]　J.B. Blair, D.L. Rabine, and M.A. Hofton, "The Laser Vegetation Imaging Sensor: a medium-altitude, digitisation-only, airborne laser altimeter for mapping vegetation and topography", *ISPRS J. Photogramm. Remote Sens.,* vol. 54, no. 2-3, pp. 115-122, 1999.

[13]　N. Haala, and C. Brenner, "Extraction of buildings and trees in urban environments", *ISPRS J. Photogramm. Remote Sens.,* vol. 54, no. 2-3, pp. 130-137, 1999.
[http://dx.doi.org/10.1016/S0924-2716(99)00010-6]

[14]　G. Patenaude, R.A. Hill, R. Milne, D.L. Gaveau, B.B. Briggs, and T.P. Dawson, "Quantifying forest above ground carbon content using LIDAR remote sensing", *Remote Sens. Environ.,* vol. 93, no. 3, pp. 368-380, 2004.
[http://dx.doi.org/10.1016/j.rse.2004.07.016]

[15]　K.S. Lim, and P.M. Treitz, "Lidar remote sensing of forest structure", *Prog. Phys. Geogr.,* vol. 27, pp. 88-106, 2003.
[http://dx.doi.org/10.1191/0309133303pp360ra]

[16]   C. Wang, and F.N. Glenn, "A linear regression method for tree canopy height estimation using airborne LIDAR data", *Can. J. Rem. Sens.,* vol. 34, pp. S217-S227, 2008.
       [http://dx.doi.org/10.5589/m08-043]

[17]   E. Næsset, "Determination of mean tree height of forest stands using airborne laser scanner data", *ISPRS J. Photogramm. Remote Sens.,* vol. 52, no. 2, pp. 49-56, 1997.
       [http://dx.doi.org/10.1016/S0924-2716(97)83000-6]

[18]   S. Magnussen, P. Eggermont, and V.N. La Riccia, "Recovering tree heights from airborne laser scanner data", *For. Sci.,* vol. 45, pp. 407-422, 1999.

[19]   J.E. Means, S.A. Acker, J. Brandon, B.J. Fritt, M. Renslow, L. Emerson, and C. Hendrix, "Predicting forest stand characteristics with airborne scanning lidar", *Photogramm. Eng. Remote Sensing,* vol. 66, pp. 1367-1371, 2000.

[20]   J. Hyyppä, O. Kelle, M. Lehikoinen, and M. Inkinen, "A segmentation-based method to retrieve stem volume estimates from 3-d tree height models produced by laser scanners", *IEEE Trans. Geosci. Remote Sens.,* vol. 39, no. 5, pp. 969-975, 2001.
       [http://dx.doi.org/10.1109/36.921414]

[21]   E. Næsset, and K-O. Bjerknes, "Estimating tree canopy heights and number of stems in young forest stands using airborne laser scanner data", *Remote Sens. Environ.,* vol. 78, pp. 328-340, 2001.
       [http://dx.doi.org/10.1016/S0034-4257(01)00228-0]

[22]   J.W. McCombs, S.D. Roberts, and D.L. Evans, "Influence of fusing LIDAR and multispectral imagery on remotely sensed estimates of stand density and mean tree height in a managed loblolly pine plantation", *For. Sci.,* vol. 49, no. 3, pp. 457-466, 2000.

[23]   S.C. Popescu, R.H. Wynne, and R.F. Nelson, "Measuring individual tree crown diameter with LIDAR and assessing its influence on estimating forest volume and biomass", *Can. J. Rem. Sens.,* vol. 29, no. 5, pp. 564-577, 2003.
       [http://dx.doi.org/10.5589/m03-027]

[24]   S.C. Popescu, R.H. Wynne, and J.A. Scrivani, "Fusion of small-footprint LIDAR and multispectral data to estimate plot-level volume and biomass in deciduous and pine forests in Virginia, USA", *For. Sci.,* vol. 50, no. 4, pp. 551-565, 2004.

[25]   R.O. Dubayah, and J.B. Drake, "LIDAR remote sensing for forestry", *J. For.,* vol. 98, no. 6, pp. 44-46, 2000.

[26]   D.J. Harding, M.A. Lefsky, G.G. Parker, and J.B. Blair, "Laser altimeter canopy height profiles: methods and validation for closed canopy, broadleaf forests", *Remote Sens. Environ.,* vol. 76, pp. 283-297, 2001.
       [http://dx.doi.org/10.1016/S0034-4257(00)00210-8]

[27]   M.A. Lefsky, W.B. Cohen, S.A. Acker, G. Parker, T. Spies, and D. Harding, "LIDAR remote sensing of the canopy structure and biophysical properties of Douglas-fir western hemlock forests", *Remote Sens. Environ.,* vol. 70, no. 3, pp. 339-361, 1999.
       [http://dx.doi.org/10.1016/S0034-4257(99)00052-8]

[28]   R. Nelson, W. Krabill, and G. Maclean, "Determining forest canopy characteristics using airborne laser data", *Remote Sens. Environ.,* vol. 15, pp. 201-212, 1984.
       [http://dx.doi.org/10.1016/0034-4257(84)90031-2]

[29]   R. Nelson, G.G. Parker, and M. Horn, "A portable airborne laser system for forest inventory", *Photogramm. Eng. Remote Sensing,* vol. 69, pp. 267-273, 2003.
       [http://dx.doi.org/10.14358/PERS.69.3.267]

[30]   R. Dubayah, and J. Drake, "LIDAR remote sensing for forestry", *J. For.,* vol. 98, pp. 44-46, 2000.

[31]   D.J. Harding, M.A. Lefsky, G.G. Parker, and J.B. Blair, "LIDAR altimeter measurements of canopy structure: Methods and validation for closed-canopy, broadleaf forests", *Remote Sens. Environ.,* vol.

76, pp. 283-297, 2001.
[http://dx.doi.org/10.1016/S0034-4257(00)00210-8]

[32]   J.C. Ritchie, D.L. Evans, D. Jacobs, J.H. Everitt, and M.A. Weltz, "Measuring canopy structure with an airborne laser altimeter", *Trans. ASAE,* vol. 36, pp. 1235-1238, 1993.
[http://dx.doi.org/10.13031/2013.28456]

[33]   E. Nasset, and T. Okland, "Estimating tree height and tree crown properties using airborne scanning laser in a boreal nature reserve", *Remote Sens. Environ.,* vol. 79, no. 1, pp. 105-115, 2012.
[http://dx.doi.org/10.1016/S0034-4257(01)00243-7]

[34]   M. Nilsson, "Estimation of tree heights and stand volume using an airborne lidar system", *Remote Sens. Environ.,* vol. 56, no. 1, pp. 1-7, 1996.
[http://dx.doi.org/10.1016/0034-4257(95)00224-3]

[35]   J. Hyyppä, and M. Inkinen, "Detecting and estimating attributes for single trees using laser scanner", *Photogrammetric Journal of Finland,* vol. 16, pp. 27-42, 1999.

[36]   P. Treitz, and K.W. Lim, "M., Pitt D, Nesbitt D, Etheridge D. LiDAR sampling density for forest resource inventories in ontario, canada", *Remote Sens.,* vol. 4, pp. 830-848, 2012.
[http://dx.doi.org/10.3390/rs4040830]

[37]   M.K. Jakubowski, Q. Guo, and M. Kelly, "Tradeoffs between lidar pulse density and forest measurement accuracy", *Remote Sens. Environ.,* vol. 130, pp. 245-253, 2013.
[http://dx.doi.org/10.1016/j.rse.2012.11.024]

[38]   P. Packalen, and M. Maltamo, "Estimation of species-specific diameter distributions using airborne laser scanning and aerial photographs", *Can J For Res-Revue Canadienne De Recherche Forestiere,* vol. 38, no. 7, pp. 1750-1760, 2008.
[http://dx.doi.org/10.1139/X08-037]

[39]   M. García, D. Riaño, E. Chuvieco, and F.M. Danson, "Estimating biomass carbon stocks for a Mediterranean forest in Spain using height and intensity LIDAR data", *Remote Sens. Environ.,* vol. 114, pp. 816-830, 2010.
[http://dx.doi.org/10.1016/j.rse.2009.11.021]

CHAPTER 6

GIS for Spatial Biology: The Geographical Component of Life

Neftalí Sillero[1,*], Cândida G. Vale[2] and Wouter Beukema[3]

[1] *CICGE, Centro de Investigação em Ciências Geo-Espaciais, Faculdade de Ciências da Universidade do Porto (FCUP), Observatório Astronómico Prof. Manuel de Barros, Alameda do Monte da Virgem, 4430-146, Vila Nova de Gaia, Portugal*

[2] *CIBIO Research Centre in Biodiversity and Genetic Resources, InBIO, Universidade do Porto, Campus Agrário de Vairão, Rua Padre Armando Quintas, Nº 7. 4485-661 Vairão, Vila do Conde, Portugal*

[3] *Department of Pathology, Bacteriology and Avian Diseases, Faculty of Veterinary Medicine, Ghent University, Salisburylaan 133, 9820, Merelbeke, Belgium*

Abstract: Spatial Biology analyses how space influences species, communities, individuals, and any other ecological processes. Geographical Information Systems (GIS) are essential in this discipline, together with remote sensing, spatial statistics, and ecological niche modelling. In this chapter, a detailed review is presented about the importance of GIS in spatial biology. Several case studies are organised at three levels, depending on the sampling unit: species, populations, and individuals. Examples are offered on species' distributions atlases; determination of chorotypes, biogeographical areas, and protected areas; modelling of species distributions, range shifts, species' dispersions, species' invasions, and hybrid zones; phylogeography and systematics; landscape connectivity; home ranges, and modelling road-kills.

Keywords: Biodiversity, Biogeography, Climate Change, Conservation, Distribution patterns, Ecological Niche Modelling, GIS, Remote Sensing, Spatial Biology, Spatial Statistics.

WHAT IS SPATIAL BIOLOGY

It is difficult to understand the living world without any reference to space, as all biological patterns and processes are related and depend on spatial characteristics, such as extension (*e.g.* area size), shape (*e.g.* orography), and distribution (*i.e.* how elements are located in the space). In fact, numerous biological questions have a spatial answer. For instance, where does a species occur? Which is its

[*] **Corresponding author Neftalí Sillero:** CICGE, Centro de Investigação em Ciências Geo-Espaciais, Faculdade de Ciências da Universidade do Porto (FCUP), Observatório Astronómico Prof. Manuel de Barros, Alameda do Monte da Virgem, 4430-146, Vila Nova de Gaia, Portugal; Tel: +351227861290; E-mail: neftali.sillero@gmail.com

Ana Cláudia Teodoro (Ed.)

distribution? Which factors delimit species' distributions? Which are the main habitats preferred by the species? Where are species affected by conservation problems? Are there other species living in the same habitats? Are there other species with similar distribution ranges? Where individuals of that species can be found? How are their home ranges composed? How do they share the space? How does competition among individuals influence space sharing? How do individuals of different species segregate? In order to answer all these questions, and various others, we need to consider space as a main object of study. How space influences species, communities, individuals, and any other ecological processes is what spatial biology analyses. This field of study is related to biogeography (the study of spatial patterns of biological diversity, *i.e.* the distribution of species and ecosystems in geographic space and through geological time [1]) and other disciplines including spatial ecology or phylogeography. By using spatial biology, it is possible to identify biogeographical and ecogeographical regions [2]; determine patterns of species distributions or chorotypes [3]; study species' ecology in order to understand the factors that restrict geographical distributions [3]; analyse species range expansions, dispersions, and invasions [4 - 6]; or identify hybrid zones among species [7, 8]. However, spatial biology focus in any biological process with spatial dimension, from global to micro-scale, and not exclusively on the ecology of species' distributions. For instance, spatial biology can analyse the distribution of diseases, of their vectors and hosts, and the factors driving them [9, 10]. Also, spatial biology may assist in making taxonomic and systematic decisions through analyses of spatial patterns of morphological and genetic trends [11]. At local scales, spatial biology can be used to investigate ecological processes related to individuals as well, including species' home ranges [12], and dispersion or migratory movements of individuals [13, 14]. In conservation biology, spatial tools can be applied to analyse problems derived from landscape fragmentation and loss of connectivity [15, 16]; wildlife road-kills [17] and risk by other human structures, like wind farms [18]; or to help to establish protected areas, like national and natural parks [19, 20].

MAIN TOOLS: GIS, REMOTE SENSING, GNSS, SPATIAL STATISTICS, ECOLOGICAL NICHE MODELS

The main tools of spatial biology are Geographical Information Systems (GIS), Remote Sensing (RS), Ecological Niche Models (ENMs), and spatial statistics. GIS can store, manipulate, analyse, and map any type of data containing spatial information, and therefore works as the main framework in which the other tools, mentioned above, are applied. With GIS support, the researcher can perform spatial statistic tests or calculate ecological niche models (in fact, a much specialised statistical test), using environmental variables obtained from RS data as main factors. All these techniques are linked and related. For instance, an ENM calculated for a species depends on the environmental variables, obtained from satellites data, but also on the species' occurrence records stored in a GIS. Further analyses of the models and the distribution pattern of the species' localities can be performed by spatial statistical tests.

GIS are currently the most applied spatial tool in spatial biology. As such, GIS have been widely used in mapping species' distributions by using cartographical capabilities [21, 22]; in identifying biogeographical regions [3, 23, 24] and species' distribution patterns [3] by statistically comparing species' maps; in inferring evolutionary scenarios [11] by using geostatistics; in determining species' connectivity through the landscape [25], looking for paths with the lowest ecological costs; in phylogeography [26] and ecological niche modelling studies [27] as main support for managing species' distribution databases and environment datasets; and in applying conservation measures [28], using data overlap tools.

RS constitutes a main source of environmental data [29, 30], mainly for very large or remote regions [3, 31], where other sources of data are not available. In fact, satellite imagery is very important for ecological niche modelling studies [31]. RS has several advantages over other data sources, notably: 1) worldwide coverage; 2) several spatial resolutions, ranging from 1 km^2 (or higher) to less than 1 m^2; 3) several temporal resolutions, ranging every several days to daily; and 4) lower costs in comparison with ground-based methods. For example, many datasets, such as altitude data from the Shuttle Radar Topography Mission (SRTM: www2.jpl.nasa.gov/srtm/) or the Landsat program (www.glovis.usgs.org), are now publicly available at no cost [30].

ENMs are empirical or mathematical estimations of the species' ecological requirements [32]. ENMs are also called species distribution, habitat distribution, or climatic envelope models [33]. ENMs are used for many purposes, including conservation management [34], to predict potentially suitable habitats for species in poorly sampled areas [35], or to investigate the effect of global warming on species' distributions [36]. ENMs can be classified into mechanistic (based on direct, experimental quantification of the effect of certain variables on species' physiology and survival) and correlative (based on statistical correlations between species occurrence records and environmental variables) models [37 - 40]. Species records can include both presences and absences, or only presences. Thus, models calculated with records including presences and absences are able to distinguish between habitats occupied and not occupied by the species, while models based on species' only-presence data provide an indication of habitat suitability, not necessarily implying that the species will be found there.

Spatial data can also be analysed statistically. Several statistical tests are implemented in GIS programmes [41], allowing the analysis of data from a complete point of view. Spatial statistical methods vary depending on the type of data to be analysed. Thus, methods analysing presence records in the form of points (or other vectorial features) are different from those analysing environmental variables in the form of raster files (images). Statistical methods for points will be focussed in looking for spatial patterns of distribution, while those used for images will be focussed in identifying, for instance, correlations among variables.

THE GIS REVOLUTION BY FREE DATA SOURCES AND APPLICATIONS

GIS are completely useless without spatial data. Obtaining spatial data of enough quality for the research purposes of spatial biology is a hard task, time-consuming, and frequently very expensive. Fieldwork is necessary to obtain these data as well as to validate them. Years ago, the only solution was that researchers produced their own data through expensive and time consuming surveys (*e.g.* producing vegetation cartography). Although satellite imagery like the produced by the Landsat program has been available since 1972, these products were not completely useful at that time. Fortunately, this situation changed profoundly when the Clinton administration (USA government) decided to eliminate the Selective Availability (*i.e.* an intentional error of around 200 m introduced in the GPS signal) on May 2, 2000, allowing users to receive a non-degraded signal globally (with an error of around 5 m). Thus, this decision opened a new and vast horizon of opportunities in civilian life. Digital cartography began to be useful as spatial positions were able to be located with precision. Google launched in 2005 Google Maps, a desktop web mapping service offering satellite imagery, street maps, 360° panoramic views of streets (Street View), real-time traffic conditions (Google Traffic), and route planning for travelling by foot, car, bicycle, or public transportation. The boost of mobile phones with integrated GPS and internet connections helped to create new applications and devices using geoinformation. These important changes in civil life also deeply influenced science. Researchers had access to new applications and new data sources. Although data validation is always strictly necessary, some cartographical products were not obtained by fieldwork anymore, but through other means like satellite images. Examples include studies depending on very high spatial resolutions, over vast areas, or in remote regions [31, 42].

At the same time the Free/Open Source Software (FOSS) movement provided a new type of license to freely use, copy, study, and change the software in any way, as the source code is openly shared so people can voluntarily improve software [30]. FOSS opened new paths for creating applications for computers at no cost for the user. There are many FOSS initiatives inside the GIS worlds, but one of the most successful is the program QGIS [43], released in 2002. This program is a cross-platform, and provides data viewing, editing, and analysis capabilities. QGIS has a small file size compared to other commercial GIS, and requires less RAM and processing power. QGIS offers some tools not available in other applications. Another very important FOSS in the academic world is R [44], a programming language and software environment for statistical computing and graphics. Currently, the R language is widely used among statisticians for developing statistical software and data analysis. Indeed, R's popularity has increased substantially in recent years. Similarly to QGIS, R allows researchers to have access to free statistical programmes with professional capabilities. The great advantage of R is its programming capacity, where users can perform very complex sequential analyses. R can also deal with spatial data, working like a GIS. Because it is a programming language, R can be incorporated to other applications. In fact, QGIS increases considerably its analytical capacities when incorporating R plugins.

CASE STUDIES

Several case studies are presented here related with spatial biology where GIS is the main tool. They are divided in three levels, depending on the sampling unit. At species level, the sampling unit is the species as a whole, with examples like distribution atlases or definition of chorotypes and biogeographic regions. At the population level, the sampling unit is the species population, *i.e.* a set of individuals of the same species. Populations allow the analyses of species' habitat requirements by ENMs, as well as dispersion processes and invasion events. In the individual level, the sampling unit is the individual of a particular species. Spatial data from individuals provide information about where they live (home ranges) or how they are affected by human structures like roads.

Species Level

Distribution Atlases (Offline and Online)

The main application of spatial biology at species level is probably to create distribution atlases [21, 22, 45], managed by a GIS. Distribution atlases are a main tool for conservation [21, 45, 46]. Conservation is not possible if what to conserve and where it occurs, is not known. Distribution atlas or ad hoc chorological information have several potential uses [21], especially in nature conservation and management, such as: 1) education and recreation, 2) documenting distribution and population, 3) documenting and analysing changes in population size and range, 4) providing a framework for survey design, 5) assessing species–

environment associations, 6) investigating theoretical aspects of ecology [47], and 7) modelling species distributions [48]. Atlases have become an indispensable data source for assessing large-scale patterns of species distributions and distributional change.

Over the years, the presentation of distribution (chorological) information has evolved from basic species lists [49, 50], to complex internet-based information visualisation methods. However, species lists only provide information about species' presences in particular areas, offering vague information about the distribution of the species [49, 50]. Distribution atlases, in contrast, offer complete information about species' chorology and are published as books [51] or articles [45], although online atlases are becoming more numerous [46]. Unfortunately, the main concern of most chorological projects is to simply store the observational data, using different ways such as ASCII files, simple spreadsheets, or databases, but mostly with no curation concerns. While databases are conclusively the most important part of a chorological project [21, 22], users attention gives greater importance to the species maps. These databases are frequently ignored by their managers upon fulfilling their goals, whereas databases should be a lasting, permanent and upgradable product, as maps are constantly out of date. Databases should allow periodic production of map compilations (published as a book, for example) in a relatively fast and simple way, but should also become a repository and tool for subsequent studies. And obviously, the best existing tool for collecting, storing, managing, mapping and analysing these databases are GIS. GPS are also essential during the process of data collection in the field, at the database creation stage, to guarantee data precision and correction. The most frequently used cartographic framework is the UTM (Universal Transverse of Mercator) grid and co-ordinate system, proposed for the first time in the Atlas Florae Europaeae (AFE) [52]. This system allows standardising surveys, to calculate hypothetical extension ranges of species, and provides a friendly display of maps.

GIS can easily build the maps as well as detect and resolve the errors included in the database. Chorological databases can be prone to numerous and varied errors, mainly in species misidentification and erroneous locality data [21]. Errors in species identification are almost impossible to correct, mainly a posteriori, if individuals' vouchers or photographs are not collected. Actually, almost all records in chorological databases only include information about the species, location, date, and collector. Similarly, geographical errors are difficult to correct but this depends on how data are collected and stored. In fact, incorrect species locations are the main source of error in chorological databases [21, 22]. The only way of avoiding them is to build secure and reliable systems where the introduction of any type of errors is reduced as much as possible through automatic data validation. For example, erroneous observational data locations can be eliminated if coordinates are recorded by GNSS enabled devices.

Determination of Chorotypes

Chorotypes are groups of species with similar distributions. Chorotypes were defined by Baroni-Urbani *et al.* [53] as clusters of species with statistically similar distributions for a specific area. Species with similar habitat requirements may have similar distributions, especially if these species were influenced by the same historical factors (*i.e.* continental drift or glacial oscillations), or present similar dispersal capacities. Distribution types are usually determined by a hierarchical cluster analysis using Jaccard's binary index and Unweighted Pair Group Method with Arithmetic mean (UPGMA) as clustering method [3, 45]. The Jaccard's index is equal to 1 when species composition is identical between squares and is equal to 0 when two squares have no species in common. According to the values of Jaccard's index, the species are clustered into a dissimilarity tree, called also dendrogram. The main chorotypes are the branches of the dendrogram with a minimum of three species and splitting off the basal polytomy of the tree.

Establishing Biogeographical Regions

The concept of biogeographical regions is related to chorotypes, and their establishment is fundamental to biogeography. Biogeographical regions are large zones of the Earth, which are generally characterised by a common geologic history, similar environmental conditions, and are composed mainly by endemic flora and fauna. Major geographical barriers (*e.g.* oceans, mountain chains, deserts) isolate each region. The ornithologist Philip L. Sclater was, in 1858, the first to propose a delimitation of terrestrial biogeographic regions based on the distributions of birds [54]. In 1879, the biologist Adolf Engler presented a division based on plant distributions. However, it was Alfred Russel Wallace (the co-author of the theory of natural selection with Charles Darwin) who set the parameters to determine the zoogeographic regions, or realms, in his classic book, The Geographical Distribution of Animals [55]. The biogeographic regions defined for flora and those created based on fauna are not similar. For instance, the South African (Capensic) region is recognised for plants but not for animals. Six realms are proposed for flora: Boreal (Holarctic), Paleotropical, Neotropical, South African (Capensic), Australian, and Antarctic. Wallace recognised six regions for fauna: Paleartic (Europe, Asia, and North Africa), Ethiopian (Africa south to Sahara desert), Oriental (India and Southeast Asia), Australian, Nearctic (North America), and Neotropical (South America). His divisions, although modified through the years, are still recognised today [2].

Determination of Protected Areas

Halting biodiversity erosion is of major priority, but also one of the largest challenges posed to current human society. To prevent biodiversity loss, it is fundamental to monitor changes in biodiversity at various spatial scales [56, 57], but also to identify areas on which conservation efforts should focus [58 - 60]. It is thus important to answer questions such as what to conserve, where best to conserve it, and how to conserve it. GIS through spatially explicit data on the ecological and geographic distribution of biodiversity can help answering the "what" and "where" questions, being particularly important to define priority areas for conservation. Furthermore, the definition and establishment of priority areas for biodiversity conservation is often severely constrained by social-economic factors [61]: resource allocation must be carefully direct to areas of highest conservation priority.

Information on the spatial distribution of biodiversity is a prerequisite for systematic conservation planning (see Distribution atlases section). In fact, the growing use of GIS prompted conservation biology studies with more robust analytical methods. However, there is still a huge lack of information since (i) much of the known species diversity has yet to be formally described and catalogued; (ii) the geographical distributions for the majority of taxa are insufficiently understood; and (iii) information on networks of interactions among organisms and physical and ecological systems remain largely deficient. Additionally, most of the available data is sparse and also skewed towards emblematic species, developed countries, regions of high accessibility, field stations, universities or museums [62, 63]. The result is that inadequate information remains available concerning most species distributions, leading to overestimated assessments and misidentification of the many priority areas for conservation.

The use of spatial data for conservation planning was introduced by Myers [64], who identified ten tropical forest "hotspots", which formed the basis for the subsequently called "Biodiversity hotspots" [59, 65 - 67]. Based on the same principals, other large-scale conservation initiatives have also attracted most of the global conservation attention, such as ecoregions [68], crisis ecoregions [69], endemic bird areas [70], mega-diverse countries, frontier forests [71], the biodiversity wilderness areas [72], roadless areas [73] and the last of the wild areas [74]. Although the application of these principals have their greatest utility at global and relatively coarse spatial scales, by helping to focus attention on broad regions of particular conservation concern, they can be also useful at finer scales [75]. In fact, a more detailed assessment is usually required within each of these regions to guide decisions about the actual location of conservation areas.

For instance, by applying high resolution spatial data and GIS to these principals, local hotspots of biodiversity were identified within areas currently considered of low priority for conservation [76]. Also rooted in the hotspots concept is the GAP analysis, which is based on the assessment of the comprehensiveness of existing protected area networks and identification of gaps in coverage [77], being applied both at regional or national scales and at global scales [78 - 80]. Notwithstanding, the latter approaches are subject to criticism, among others, regarding excluding economic and social factors [81] and, for not establishing conservation goals [82].

In the last decades, a new field of 'systematic' conservation planning has emerged [83]. It was enhanced by the emergence of quantitative tools to address spatial prioritisation problems and to guide the decision making process [84]. Systematic conservation planning developed a framework to efficiently identify priority areas for conservation that maximises representation and persistence, in other words, must ensure that these areas are sufficiently large, well connected and well replicated to promote long-term persistence of the biodiversity encompassed [61, 85, 86]. This approach requires spatially explicit data on biodiversity distribution (species, communities, functional and/or genetic diversity) [83, 87, 88], as well as, the ecological and evolutionary determinants of its spatial patterns. Generally, systematic conservation planning addresses two major classes of conservation prioritisation problems: the "minimum-set" and the "maximum-cover" (see [89 - 91] for reviews). The first problem aims to identify the minimum number of sites, or minimum total area, that capture a set amount of biodiversity for the least cost [92 - 94]. While, the "maximum-cover" problem aims to find an area that captures as much biodiversity as possible below a fixed cost [95 - 97]. By using heuristics, meta-heuristics and optimal algorithms, these problems can be formalised and solved mathematically using computational tools, such as Marxan [98] and Zonation [84]. The latter are two of the most widely used conservation-planning, decision-support tools. Marxan uses a minimum-set approach to identify planning units that achieve conservation targets at near-minimal cost [98]. While, Zonation uses a maximum-cover approach that aims to maximize the conservation benefits for a fixed cost specified by the user [84]. Some studies have been already applying these tools to design proactive conservation plans under different assumptions. For instance, a modified version of Marxan was used to study a method to account for uncertainty in species distributions under climate change and make trade-offs in investment options [19]. Other approaches with Marxan incorporate evolutionary processes into conservation planning using species distribution patterns and environmental gradients as surrogates for genetic diversity [99] or identification of priority areas for conservation in remote areas [100]. Examples using Zonation are for instance in studies for assessing the habitat connectivity and allocating management actions [101] or to enhance biological diversity protection while promoting sustainable development and providing spatial guidance in the resolution of potential policy conflicts over priority areas for conservation at risk of transformation [102].

Population Level

Ecological Niche Models

The purpose of ENMs is to provide spatially explicit information on species, communities and other elements of biodiversity for conservation planning, risk assessment and resource management. The goal of ENMs is to derive models of environmental suitability for species in space and time. This is achieved by identifying statistical relationships between spatial data of species observations (presence or abundance data) and of environmental descriptors [38, 103 - 105]. In this context, several algorithms have been either developed or applied to estimate distributional areas on the basis of correlations of known species occurrences with environmental variables. Depending on the question or type of data, available algorithms range from very simplistic envelope models (*i.e.* envelope models define minimum and maximum values of climate boundaries around species occurrences, thereby delimiting a 'climate envelope' within which species occur) to regression based analyses, and the more complex machine-learning techniques (see [37, 106] for reviews).

Although several modelling techniques are now available for identifying distributional patterns, ENMs have numerous assumptions being subject to multiple sources of uncertainty. Among others, uncertainties might derived from sampling design, the selection of environmental factors, modelling techniques applied or scale related issues [27,107–111]. Thus, fitting ENMs requires numerous methodological and well-justified decisions. Notwithstanding, ENMs have become increasingly popular in recent years at the point that examples of their applications are widespread and diverse. The most common application is the identification of current suitable areas for species occurrence [37], either for rare species [112] or over large and/or remote study areas [3, 31, 113]; assessing the conservation status of poorly known species [114, 115]; predicting species distribution both in past and future time [116], identify hotspots of richness [3] and identification of priority areas for conservation [19, 100]. ENMs are thus the base for most of the subjects discussed below, such as, species' range shifts. Indeed, advances in computing science and the development of large datasets have made ENMs easier, faster and thus more widely applied to several fields including quantitative ecological studies [117], evolutionary biology [118], population genetics [119], landscape genetics [120], biogeography [121], climate change [36, 122], and conservation [83, 105, 123].

Modelling Species' Range Shifts

Under the current biodiversity crises, modelling species range shifts is crucial. It helps answering questions such as: Will species survive to future climate change? Will they lose or gain suitable areas? Will they be able to reach new suitable areas? Will species' new suitable areas be covered by current protected areas? These and other questions are fundamental in biodiversity conservation studies, particularly for designing proactive conservation plans. GIS, spatial data and ENMs are essential to predict species' range shifts across space and time. However, ENMs often fail in incorporating factors such as species dispersal capacity and/or potential for rapid evolutionary adaptation [124, 125].

In terrestrial systems, climate dominates the distribution of species at global scale. Distributional limits are often correlated with particular combinations of climate variables and habitat factors, changing through time as the climate itself changes [126]. Examples include latitudinal shifts in species' distributions in response to past climatic oscillations [127] or recent poleward expansions of plants [128], breeding birds [129] or butterflies [130], as well as elevation and latitudinal shifts of vertebrates in response to anthropogenic climate change [122, 131 - 134].

Species' potential range shifts responses to future climate changes are primarily driven by their level of exposure, followed by their sensitivity and adaptive capacity [135, 136]. Under future climate scenarios, species distributions might remain stable, disappear or even increase [137]. Under increasing temperatures, some climate regimes are expected to shift upward in elevation and poleward in latitude [130]. Lowlands are expected to be more vulnerable to future climate change than montane areas, because high velocities of climate change are expected in lowlands [138].

Future climate change' assessments require a species-by-species analysis that incorporates biological traits, combined with analyses of the effects of different dimensions of climate change. Recent examples include for instance assessing the impacts of future climate change in plant–insect interactions [139] or forecasted future distributional changes in communities of plants and birds [140]. Other applications are directly related with the incorporation of the predicted responses of species to future climate change in systematic conservation planning [99, 141].

Dispersion Models

The capacity of a species to disperse is one of the main factors that determine its potential niche [142]. While incorporating dispersal is of major importance when building an ENM, only a restricted number of ecological niche modelling studies have attempted to do so during the last decade (but see *e.g.* [132, 143].). A lack of basic natural history information on the species often underlies this issue. Failing to correct models for dispersal limitations may affect accuracy, chiefly due to overprediction, in which the presence of a species is predicted in an area that it in reality cannot reach [132]. In turn, these errors can influence policy decisions regarding conservation or projected area planning [144]. During recent years, dispersal ability has been increasingly used, often in climate change studies which focus on the potential of species to persist in a changing environment [145 - 147]. GIS is indispensable in this sense, as it combines the different variables (*e.g.* land use, abiotic barriers, competing species distributions) that could influence dispersal. Various GIS techniques have been developed that incorporate dispersal using various scenarios (*e.g.* unlimited, restricted or no dispersal), including BioMove [148] and the MIGCLIM in R [149]. Various other GIS approaches are available, depending on the preferred platform used.

Modelling Species Invasions

In addition to destruction and fragmentation of habitats and climate change, biological invasions have also been suggested to drive species extinctions [150]. The rates of these processes, and in particular of biological invasion, have strongly increased during the last century due to human growth and globalisation causing a parallel increase in the rate of biodiversity loss: the negative impact of species' introductions can unbalance the native ecosystem structure through competition, predation, genetic contamination, spread of diseases, and complex modifications of habitats [150]. GIS tools are essential to analyse expansion patters of invasive species. Particularly, ENMs are useful to define the environmental preferences and the potential invasiveness of invasive species [5, 151], to predict their possible expansion routes, and also to identify areas which are more likely to be prone to invasions currently [152, 153] or under scenarios of future climate change [151].

Modelling Species Hybrid Zones

The development of GIS enhanced the understanding of the spatial component of evolutionary processes. Indeed, spatial patterns are the main explanation for several evolutionary hypotheses. In the field of phylogeography, several studies have applied GIS-based models to spatially analyse the influence of environmental factors in maintenance hybrid zones for various species [7]. The major focus for studying hybrid zones has been the relative importance of exogenous factors in determining the location and maintenance of hybrid zones [154]. Due to the obvious spatial component of this enquiry, GIS and ENMs are fundamental tools in hybrid zone research. Particularly, ENMs are useful for understanding and identifying the environmental variables promoting ecological divergence. In the last decades, a growing number of studies have applied this framework, but the very first study dates back to 1988. By using the ENM method BIOCLIM, Kohlman *et al.* [155] tested if extrinsic forces could explain the location of parapatric distributional boundaries among chromosomal races within an Australian grasshopper. Since this study, many others have applied different ENMs to different questions in hybrid zones [7, 156 - 159]. Despite being widely used, most of these studies are based on models that rely on presence-only data [7] or require the reclassification of continuous data into discrete entities, by using genetic information for limiting populations, lineages or even species [156]. This limits models' power to detect niche divergence between hybridising taxa. Modelling both pure and hybrid individuals (considering the full gradient from one taxon to the other) enhance the models ability to assess the ecological agents of selection that are relevant for species formation. Therefore, rather than using presence-only data or assigning individuals to a given species or cluster, using the cluster membership probability determined by genetic analysis as input data is a more efficient solution [160]. This approach has several advantages including the ability to produce detailed ecological models directly from population structure, taking into account the full genetic continuum characterising hybrid zones; to define hybrid zones without requiring a model based on the locations of the hybrid individuals; and working at local scales [160].

Linking Phylogeography with Ecological Niche Models

While phylogenetic analyses assess relationships between species or other taxonomical groups, ENMs and other GIS-related methods have proven to be extremely useful in explaining their recovered patterns or relationships (see [161, 162] for review). Many of the fundamental questions in phylogeography relate to how physical and climatic variation over space and time shapes patterns of genetic divergence; the relevance of these abiotic factors can be tested in a GIS. Early studies in this field created their own spatial variables based on temperature- and precipitation data, followed by statistical analyses to test hypotheses on phylogenetic relationships [118]. Nowadays, a similar approach is followed, although various freely-available datasets of contemporary, historical and predicted future environmental conditions [163] make it possible to test hypotheses in a wide variety of approaches developed specifically for this purpose. For instance, Waltari *et al.* [164] combined a phylogeography with niche models which they hindcasted onto historical climate data (creating paleo-distribution models) to identify the location of refugia (*i.e.* climatic refuges during ice ages) of 20 North American terrestrial vertebrate species. Carstens and Richards [165] applied a similar approach to show that current, congruent phylogeographic patterns can arise from incongruous ancestral distributions, using vertebrates and plants as model species.

Other studies at species level combining phylogeographies and ENMs have focused on making systematic decisions within species complexes (see next section). Also, ENMs together with genetic and demographic models studied how distributional shifts might leave species-specific signatures in patterns of genetic variation [166]. Yet, the combination of these GIS approaches is also usable above or below species level. Banta *et al.* [167] modelled the climatic niche of various genotypes of a plant to test whether phenotypic variation of ecologically important traits in that species was spatially structured. By using a holistic approach, Edwards *et al.* [168] explored whether genetic divergence among an Australian reptile community could be explained by biogeographic vicariance across barriers, habitat stability, population isolation along a linear habitat or fragmentation across different environments.

Ecological Niche Models as a Tool for Systematics

What exactly a species is, and how to define it, remains a topic of discussion. Nevertheless, recent years have seen a trend towards the use of integrative data to delineate species [169]. GIS and ENMs have been applied as part of integrative analyses to test whether different (candidate) species present different environmental preferences. In this sense, niches of two or more species are described by means of ENM, based on data prepared in a GIS environment, and further compared using various spatio-statistical tests to assess whether and how they differ [170, 171]. Whereas the use of spatial data in this sense seems promising, several issues remain. For instance, although a clear definition of subspecies (the unit below species level) is lacking, a consensus exists that such taxa possess slightly different environmental preferences, being therefore separated, but forming broad contact zones with other subspecies [172]. What follows is that a certain level of environmental variation usually exists within wide-ranging, polytypic species, due to which the combination of a GIS with for instance molecular data (to *e.g.* assess the degree gene-flow) is indispensable to make systematic decisions [173 - 175]. Incorporation of GIS-related data in systematic studies should only be done in combination with other data, and when sampling for instance species' occurrence localities for an ENM, the entire species' range must be considered. While doing so, the ENMs serve two purposes: they act as an independent line of evidence, but can also generate species hypotheses to be explored in greater detail with additional fieldwork and molecular approaches. The recent use of physiological data such as water loss or temperature preferences as part of spatial mechanistic models can greatly inform decisions making in systematic studies [176].

Landscape Connectivity

Habitat loss, fragmentation, and degradation are major direct causes of threats to biodiversity [177]. Accessing suitable habitats is essential to species survival: species' populations become extinct when their habitat disappears. However, habitat fragmentation and degradation are more common in comparison to total disappearance of a habitat. Habitat fragmentation is the division of a habitat in smaller patches, isolated from each other by a matrix of habitats different from the initial one [177]. Isolation of habitat patches affects the general connectivity of regions, changing species' daily movements, dispersal patterns, and home ranges [178]. Consequently, gene flow among populations can be interrupted, leading to a reduction of genetic diversity and to an increment of extinction proneness. On the other hand, habitat degradation is the loss of habitat quality [178]. Normally, it is a gradual process along time affecting mainly food and shelter availability.

Landscape connectivity is the amount of permeability among habitat patches. The more permeable a habitat is, the higher is the connectivity. Animals' movements among patches can be facilitated if the landscape connectivity is higher. Connectivity includes structural connectivity, *i.e.* the physical characteristics of the landscape, and functional connectivity, *i.e.* the actual flux of organisms through the landscape, which depends on the behavioural response of the organisms [25].

The community connectivity can be analysed from a landscape point of view [25, 179], applying graph theory framework [180] with the help of a GIS, to relate, for instance, the connectivity importance with the species richness. Briefly, a set of discrete habitats (usually favourable habitat patches) can be represented in a graph as nodes connected with each other by edges: the linkages between the nodes implying connectivity [181, 182]. In some cases, network connectivity may influence the species richness pattern [25].

Landscape connectivity can be represented by a permeability matrix or friction map, where permeability is defined as the energy cost of moving from one place to another [183]. Therefore, the friction map is composed by a matrix (*i.e.* in raster format) of suitable and unsuitable habitat patches, where each cell is weighted by a value that represents its permeability. The highest permeability is assigned to the most favourable variable value for the species movements; the lowest permeability represents those habitats hampering species movements. The final friction map is the addition of all variables, giving a final value per pixel meaning the total permeability of that pixel.

Individual Level

Home Ranges

At the individual level, spatial biology studies species' home ranges, *i.e.* the basic unit of the species' distributions [184]. Using mammals as models, Burt [185], defined home range as the "area traversed by the individual in its normal activities of food gathering, mating and caring for young. Occasional sallies outside the area, perhaps exploratory in nature, should not be considered as in part of the home range". Thus, the home range of an individual is the spatial representation of the distribution of all the items needed for the survival of that individual [185]. In this sense, the home range is the cognitive map of the individual, built by the animal's understanding of the surrounding environment [12].

Home ranges have several spatial characteristics (size, shape, and structure) which are affected by factors such as animal size, metabolic needs, resource availability, and population density [186]. However, topography may constrain the size and shape of home ranges by making its limits coincide with particular topographic features [187 - 189].

GIS provide many different tools to calculate the home range: minimum convex polygon, characteristic hull polygons, and kernel density functions [190]. Moreover, GIS are able to analyse how the home ranges of several individuals overlap in space and time, through vectorial analysis tools. It is also possible to identify the distribution patterns of a group of individuals using clustering statistics. For instance, researchers are interested to understand how individuals of two species share the same space [191]; do the species segregate in space or in time? Is it possible to identify clusters of individuals or they are distributed homogeneously across the area? In this context, it is very important to find out how species abundance is distributed and by which limiting factors this remains restricted [192].

At individual level, researchers are constrained by the availability of environmental data of high spatial resolution. RS imagery provides an optimal solution: either through satellite images, aerial photographies, or LiDAR data, available at spatial resolutions lower than 1 m [42]. New devices like drones allow recording environmental variables (altitude, vegetation, temperature) for very small areas with a very high spatial resolution at low costs. These approaches will make it possible to analyse the home range of small-sized species, taking into account the spatial properties of the habitat.

Distribution Patterns of Species Communities

Analyses of spatial patterns are essential in ecology and provide insights about how organisms interact among each other or with their environment [193]. The analysis of the spatial structure of community composed by several species allows understanding how individuals share the space, and modify their home ranges depending on the distribution of energy sources (light, water, soil nutrients, food), availability of shelters and resting places, presence of other species (predators, competitors, disease vectors), mates and conspecifics. While home range analyses refer to the spatial needs of each individual, ENMs focus on the presence of one or more species in a particular geographical area characterised by certain environmental variables (mainly climate) at coarse scales (regional or continental). Linking both analytical levels, community spatial distributions can identify the main environmental factors with a high spatial resolution, applying techniques of species distribution studies to the home range scale. In this sense, species' interactions can be more easily analysed without collecting repeatedly individual data (as required for home range studies).

Species within a community can be distributed randomly, regularly, or in clusters [194, 195]. Random distributions are typical of non-territorial or non-competing species living in habitats with abundant and widespread resources. The probability to find an individual is the same everywhere and independent from the presence of other individuals. Species can be regularly distributed when individuals avoid being mutually close (because of competition or strong territoriality) and resources are evenly distributed. Species appear clustered when the resources are irregularly distributed, hence, the probability to find either a second individual near the first or areas without individuals is higher than expected by random. Clustering is the most frequent pattern observed in species communities [196] with more or less intensity [197]. Some species change from regular to clustered distribution when density increases [194].

The analyses of spatial patterns are performed using spatial statistics [198, 199]. The most frequently used tests in the analysis of distribution patterns are related to point processes: Ripley's K function, to determine the distribution pattern along distances [200]; the Nearest Neighbour Index, to measure the degree of aggregation of a point process [201]; the Moran's I index [202], for global autocorrelation, *i.e.* the correlation of a point process in relationship with a particular variable; and Local Indicators of Spatial Association (LISA) [203], for local autocorrelation, in order to identify internal spatial patterns of autocorrelation. These statistical tests help identify how the individuals are distributed, their relationship with the spatial factors (*i.e.* the degree of autocorrelation with the resources), and the local patterns of autocorrelation and outliers. For instance, tree species tend to be clustered, but dead individuals are randomly distributed [204]. This pattern can change with age, from clustered distribution in younger plants to random in median-aged, and then to regular in the oldest ones, either in trees [205, 206] or shrubs [207, 208]. The main factor driving the plants' distribution patterns is competition [204, 207], but it can be also substrate [209]. In animals, there are fewer studies describing how species are locally distributed. The most frequent pattern is clustered [196], with more or less intensity [197]. Some species change from regular to clustered distribution when density increases [194]. Adults can be randomly distributed, while juveniles have frequently a regular distribution, as they are excluded to less suitable habitats [42, 195].

Modelling Road-kills

Roads have various effects on wildlife [210 - 213]: 1) mortality from road construction and vehicle collisions; 2) habitat loss and fragmentation by alteration of the physical environment; 3) subdivision of populations into smaller and more vulnerable units; 4) interference with animal behaviour, changing home ranges, movement, reproductive success, escape response, and physiological state; and 5) spread of exotic species promoted by habitat alteration, stressing native species, and providing movement corridors. All these effects can extend from 100 m to 1 km around the road itself [214]. Road-kills are the greatest source of direct human-induced terrestrial wildlife mortality, with significant impacts at the population level [215 - 217]. The distribution of road-kills is usually not random, because animals frequently cross roads using the same routes [218 - 221]. Thus, it is possible to analyse the distribution of road-kills, to identify the main migration routes, and then to define road-kill hotspots where conservation measures should be implemented. All these analyses can be performed using GIS and spatial statistics.

The Nearest Neighbour Index (NNI) can be used to determine if species records on roads are clustered or not [201]. The NNI is expressed as the ratio of the observed distance between points and the expected distance, *i.e.* the average distance between neighbours in a hypothetical random distribution. If the index is less than 1, records are clustered; if the index is greater than 1, records are dispersed. The NNI is usually calculated for road-killed and alive specimens separately, in order to examine whether the spatial distribution of both groups qualitatively differ or not. Both groups can have a different distribution pattern when road-kills are determined by some factors such as traffic intensity.

Sometimes, road-kills appear at very high proportions, the so-called hotspots. By using spatial statistics, the spatial pattern of road-kills is compared with a random model where the likelihood of road-kills for each road section follows a Poisson distribution [218]. Using this hypothesis and given the mean number of road-kills per road section, it is possible to calculate the probability of any road section having x number of road-kills. In a random distribution of road-killed specimens, the hotspots of road-kills should be distributed at random across the roads, and their aggregation would be extremely unlikely.

Animals' crossing routes are determined by many factors, namely, habitat composition or structure and road texture, among others. Small and linear habitats can enhance or limit dispersal movements, depending if they act as corridors or barriers [183]. Animals usually move through routes with the lowest energy costs for dispersal movements. Then, crossing routes can be identified using a permeability matrix or friction map [183].

CONCLUSIONS

Life is influenced by space. Therefore, to understand life, we need to analyse it from a spatial point of view. GIS spatial statistics, ENMs, and RS data are the main tools used for such task. However, from a technical point of view, spatial research is not complete. Researchers should understand the importance of good quality of spatial data. In this sense, all species' records should be recorded with a GNSS receptor in order to avoid errors in geographical location, one of the main problems in chorological databases [21]. Moreover, research on obtaining new environmental variables from satellite imagery is necessary. Applied investigation on the use of non-piloted vehicles (*i.e.* drones) is essential in order to allow the recording of environmental variables (altitude, vegetation, temperature) for very small areas using high spatial resolution at lower costs. Spatial statistics should be widely implemented in GIS software [41], which constitutes probably the most important deficiency in GIS-related analyses. Finally, several factors affecting the behaviour of ecological niche models should be investigated, like the effect of shape and size of the study area. New statistical tests should be implemented in order to measure the quality and accuracy of ENMs [222]. And perhaps more important, researchers should understand what ENMs does model [33]. Altogether, these recommendations will result in a better use of GIS and associated techniques, to improve our understanding of life on Earth.

CONSENT FOR PUBLICATION

Not applicable.

CONFLICT OF INTEREST

The author (editor) declares no conflict of interest, financial or otherwise.

ACKNOWLEDGEMENTS

NS is supported by a research contract (IF/01526/2013) from Fundação para a Ciência e Tecnologia (FCT, Portugal).

REFERENCES

[1]   M.V. Lomolino, B.R. Riddle, R.J. Whittaker, and J.H. Brown, *Biogeography.* Sinauer Associates: Massachusetts, 2005.

[2]   B.G. Holt, J-P. Lessard, M.K. Borregaard, S.A. Fritz, M.B. Araújo, D. Dimitrov, P.H. Fabre, C.H. Graham, G.R. Graves, K.A. Jønsson, D. Nogués-Bravo, Z. Wang, R.J. Whittaker, J. Fjeldså, and C. Rahbek, "An update of Wallace's zoogeographic regions of the world", *Science,* vol. 339, no. 6115, pp. 74-78, 2013.
[http://dx.doi.org/10.1126/science.1228282] [PMID: 23258408]

[3]   N. Sillero, A.K. Skidmore, A.G. Toxopeus, and J.C. Brito, "Biogeographical patterns derived from remote sensing variables: the amphibians and reptiles of the Iberian Peninsula", *Amphib.-Reptil.,* vol. 30, no. 2, pp. 185-206, 2009.
[http://dx.doi.org/10.1163/156853809788201207]

[4]   G.F. Ficetola, W. Thuiller, and C. Miaud, "Prediction and validation of the potential global distribution of a problematic alien invasive species: the American bullfrog", *Divers. Distrib.,* vol. 13, no. 4, pp. 476-485, 2007.
[http://dx.doi.org/10.1111/j.1472-4642.2007.00377.x]

[5]   G.F. Ficetola, W. Thuiller, and E. Padoa-Schioppa, "From introduction to the establishment of alien species: bioclimatic differences between presence and reproduction localities in the slider turtle", *Divers. Distrib.,* vol. 1, no. 15, pp. 108-116, 2009.
[http://dx.doi.org/10.1111/j.1472-4642.2008.00516.x]

[6]   I. Silva-Rocha, D. Salvi, N. Sillero, J.A. Mateo, and M.A. Carretero, "Snakes on the Balearic islands: an invasion tale with implications for native biodiversity conservation", *PLoS One,* vol. 10, no. 4, p. e0121026, 2015.
[http://dx.doi.org/10.1371/journal.pone.0121026] [PMID: 25853711]

[7]   F. Martínez-Freiría, N. Sillero, M. Lizana, and J.C. Brito, "GIS-based niche models identify environmental correlates sustaining a contact zone between three species of European vipers", *Divers. Distrib.,* vol. 14, no. 3, pp. 452-461, 2008.
[http://dx.doi.org/10.1111/j.1472-4642.2007.00446.x]

[8]   B. Wielstra, W. Babik, and J.W. Arntzen, "The crested newt Triturus cristatus recolonized temperate Eurasia from an extra-Mediterranean glacial refugium", *Biol. J. Linn. Soc. Lond.,* vol. 114, pp. 574-587, 2015.
[http://dx.doi.org/10.1111/bij.12446]

[9]   A.H. Dicko, R. Lancelot, M.T. Seck, L. Guerrini, B. Sall, M. Lo, M.J. Vreysen, T. Lefrançois, W.M. Fonta, S.L. Peck, and J. Bouyer, "Using species distribution models to optimize vector control in the framework of the tsetse eradication campaign in Senegal", *Proc. Natl. Acad. Sci. USA,* vol. 111, no. 28, pp. 10149-10154, 2014.
[http://dx.doi.org/10.1073/pnas.1407773111] [PMID: 24982143]

[10]   Y. Si, W.F. de Boer, and P. Gong, "Different environmental drivers of highly pathogenic avian influenza H5N1 outbreaks in poultry and wild birds", *PLoS One,* vol. 8, no. 1, p. e53362, 2013.
[http://dx.doi.org/10.1371/journal.pone.0053362] [PMID: 23308201]

[11]   J.C. Brito, X. Santos, J.M. Pleguezuelos, and N. Sillero, "Inferring Evolutionary Scenarios with Geostatistics and Geographical Information Systems (GIS) for the viperid snakes Vipera latastei and V. monticola", *Biol. J. Linn. Soc. Lond.,* vol. 95, pp. 790-806, 2008.
[http://dx.doi.org/10.1111/j.1095-8312.2008.01071.x]

[12]   R.A. Powell, and M.S. Mitchell, "What is a home range?", *J. Mammal.,* vol. 93, no. 4, pp. 948-958, 2012.
[http://dx.doi.org/10.1644/11-MAMM-S-177.1]

[13]   C. Curtice, D.W. Johnston, H. Ducklow, N. Gales, P.N. Halpin, and A.S. Friedlaender, "Modeling the spatial and temporal dynamics of foraging movements of humpback whales (Megaptera novaeangliae) in the Western Antarctic Peninsula", *Mov. Ecol.,* vol. 3, no. 1, p. 13, 2015.
[http://dx.doi.org/10.1186/s40462-015-0041-x] [PMID: 26034604]

[14]   M. Franch, A. Montori, N. Sillero, and G.A. Llorente, "Temporal analysis of Mauremys leprosa (Testudines, Geoemydidae) distribution in northeastern Iberia: unusual increase in the distribution of a native species", *Hydrobiologia,* vol. 757, no. 1, pp. 129-142, 2015.
[http://dx.doi.org/10.1007/s10750-015-2247-8]

[15]   S. Saura, and J. Torné, "Conefor Sensinode 2.2: A software package for quantifying the importance of habitat patches for landscape connectivity", *Environ. Model. Softw.,* vol. 24, no. 1, pp. 135-139, 2009.
[http://dx.doi.org/10.1016/j.envsoft.2008.05.005]

[16]   G. Baranyi, S. Saura, J. Podani, and F. Jordán, "Contribution of habitat patches to network connectivity: Redundancy and uniqueness of topological indices", *Ecol. Indic.,* vol. 11, no. 5, pp. 1301-1310, 2011.
[http://dx.doi.org/10.1016/j.ecolind.2011.02.003]

[17]   C. Matos, N. Sillero, and E. Argaña, "Spatial analysis of amphibian road mortality levels in northern Portugal country roads", *Amphib.-Reptil.,* vol. 33, pp. 469-483, 2012.
[http://dx.doi.org/10.1163/15685381-00002850]

[18]   H. Santos, L. Rodrigues, G. Jones, and H. Rebelo, "Using species distribution modelling to predict bat fatality risk at wind farms", *Biol. Conserv.,* vol. 157, pp. 178-186, 2013.
[http://dx.doi.org/10.1016/j.biocon.2012.06.017]

[19]   S.B. Carvalho, J.C. Brito, E.J. Crespo, and H.P. Possingham, "Incorporating evolutionary processes into conservation planning using species distribution data: a case study with the western Mediterranean herpetofauna", *Divers. Distrib.,* vol. 17, no. 3, pp. 408-421, 2011.
[http://dx.doi.org/10.1111/j.1472-4642.2011.00752.x]

[20]   P. De Pous, E. Mora, M. Metallinou, D. Escoriza, M. Comas, and D. Donaire, "Elusive but widespread? The potential distribution and genetic variation of Hyalosaurus koellikeri (Günther, 1873) in the Maghreb", *Amphib.-Reptil.,* vol. 32, pp. 385-397, 2011.
[http://dx.doi.org/10.1163/017353711X587732]

[21]   N. Sillero, L. Celaya, and S. Martín-Alfageme, "Using GIS to Make An Atlas: A Proposal to Collect, Store, Map and Analyse Chorological Data for Herpetofauna", *Rev. Esp. Herpetol.,* vol. 19, pp. 87-101, 2005.

[22]   A. Loureiro, and N. Sillero, Metodologia.*Atlas dos anfíbios e répteis de Portugal.,* A. Loureiro, N. Ferrand, M.A. Carretero, O. Paulo, Eds., Esfera do Caos: Lisboa, 2010, pp. 66-74.

[23]   J.M. Vargas, R. Real, and J.C. Guerrero, "Biogeographical regions of the Iberian peninsula based on freshwater fish and amphibian distributions", *Ecography,* vol. 4, no. 21, pp. 371-382, 1998.
[http://dx.doi.org/10.1111/j.1600-0587.1998.tb00402.x]

[24]   M.A. Vidal, E.R. Soto, and A. Veloso, "Biogeography of Chilean herpetofauna: distributional patterns of species richness and endemism", *Amphib.-Reptil.,* vol. 30, pp. 151-171, 2009.
[http://dx.doi.org/10.1163/156853809788201108]

[25]   R. Ribeiro, N. Sillero, M.A. Carretero, G. Alarcos, M. Ortiz-Santaliestra, and M. Lizana, "The pond network: Can structural connectivity reflect on (amphibian) biodiversity patterns?", *Landsc. Ecol.,* vol. 26, pp. 673-682, 2011.
[http://dx.doi.org/10.1007/s10980-011-9592-4]

[26]   A.W. Hill, and R.P. Guralnick, "GeoPhylo: an online tool for developing visualizations of

phylogenetic trees in geographic space", *Ecography,* no. 33, pp. 633-636, 2010.
[http://dx.doi.org/10.1111/j.1600-0587.2010.06312.x]

[27]   M.A. Barbosa, N. Sillero, F. Martínez-Freiría, and R. Real, *Ecological niche models in Mediterranean herpetology: past, present and future.* Ecological Modeling. WenJun Zhang, 2012, pp. 173-204.

[28]   J.M. Benayas, Ede.L. Montaña, J. Belliure, and X.R. Eekhout, "Identifying areas of high herpetofauna diversity that are threatened by planned infrastructure projects in Spain", *J. Environ. Manage.,* vol. 79, no. 3, pp. 279-289, 2006.
[http://dx.doi.org/10.1016/j.jenvman.2005.07.006] [PMID: 16253418]

[29]   J.T. Kerr, and M. Ostrovsky, "From space to species: ecological applications for remote sensing", *Trends Ecol. Evol.,* vol. 18, no. 6, pp. 299-305, 2003.
[http://dx.doi.org/10.1016/S0169-5347(03)00071-5]

[30]   N. Sillero, and P. Tarroso, "Free GIS for herpetologists: free data sources on Internet and comparison analysis of proprietary and free/open source software", *Acta Herpetol.,* vol. 5, no. 1, pp. 63-85, 2010.

[31]   N. Sillero, J.C. Brito, S. Martín-Alfageme, E. García-Meléndez, A.G. Toxopeus, and A. Skidmore, "The significance of using satellite imagery data only in Ecological Niche Modelling", *Acta Herpetol.,* vol. 7, no. 2, pp. 221-237, 2012.

[32]   N. Sillero, A.M. Barbosa, F. Martínez-Freiría, and R. Real, "Los modelos de nicho ecológico en la herpetología ibérica: pasado, presente y futuro", *Bol la Asoc Herpetol Esp,* vol. 21, pp. 2-24, 2010.

[33]   N. Sillero, "What does ecological modelling model? A proposed classification of ecological niche models based on their underlying methods", *Ecol. Modell.,* vol. 222, pp. 1343-1346, 2011.
[http://dx.doi.org/10.1016/j.ecolmodel.2011.01.018]

[34]   L. Bulluck, E. Fleishman, C. Betrus, and R. Blair, "Spatial and temporal variations in species occurrence rate affect the accuracy of occurrence models", *Glob. Ecol. Biogeogr.,* vol. 15, no. 1, pp. 27-38, 2006.
[http://dx.doi.org/10.1111/j.1466-822X.2006.00170.x]

[35]   R. Engler, A. Guisan, and L. Rechsteiner, "An improved approach for predicting the distribution of rare and endangered species from occurrence and pseudo-absence data", *J. Appl. Ecol.,* vol. 41, pp. 263-274, 2004.
[http://dx.doi.org/10.1111/j.0021-8901.2004.00881.x]

[36]   S.B. Carvalho, J.C. Brito, E.J. Crespo, and H.P. Possingham, "From climate change predictions to actions- conserving vulnerable animal groups in hotspots at a regional scale", *Glob. Change Biol.,* vol. 16, no. 12, pp. 3257-3270, 2010.
[http://dx.doi.org/10.1111/j.1365-2486.2010.02212.x]

[37]   A. Guisan, and N.E. Zimmermann, "Predictive habitat distribution models in ecology", *Ecol. Modell.,* vol. 135, pp. 147-186, 2000.
[http://dx.doi.org/10.1016/S0304-3800(00)00354-9]

[38]   A. Guisan, and W. Thuiller, "Predicting species distribution: offering more than simple habitat models", *Ecol. Lett.,* vol. 9, no. 8, pp. 993-1009, 2005.
[http://dx.doi.org/10.1111/j.1461-0248.2005.00792.x]

[39]   A.T. Peterson, "Uses and Requirements of Ecological Niche Models and Related Distributional Models", *Biodivers. Inform.,* no. 3, pp. 59-72, 2006.

[40]   M. Kearney, and W. Porter, "Mechanistic niche modelling: combining physiological and spatial data to predict species' ranges", *Ecol. Lett.,* vol. 12, no. 4, pp. 334-350, 2009.
[http://dx.doi.org/10.1111/j.1461-0248.2008.01277.x] [PMID: 19292794]

[41]   L. Anselin, I. Syabri, and Y. Youngihn Kho, "GeoDa: An Introduction to Spatial Data Analysis", *Geogr. Anal.,* vol. 38, no. 1, pp. 5-22, 2006.
[http://dx.doi.org/10.1111/j.0016-7363.2005.00671.x]

[42]   N. Sillero, and L. Gonçalves-Seco, "Spatial structure analysis of a reptile community with airborne LiDAR data", *Int. J. Geogr. Inf. Sci.,* vol. 28, no. 8, pp. 1709-1722, 2014.
[http://dx.doi.org/10.1080/13658816.2014.902062]

[43]   QGIS Development Team, "QGIS Geographic Information System", *Chicago: Open Source Geospatial Foundation.* http: //qgis.osgeo.org

[44]   R Development Core Team, "A language and environment for statistical computing", *Vienna: R Foundation for Statistical Computing.* http: //www.r-project.org/

[45]   N. Sillero, A. Bonardi, C. Corti, R. Creemers, P. Crochet, and G.F. Ficetola, "Updated distribution and biogeography of amphibians and reptiles of Europe", *Amphib.-Reptil.,* vol. 35, no. 1, pp. 1-31, 2014.
[http://dx.doi.org/10.1163/15685381-00002935]

[46]   N. Sillero, M.A. Oliveira, P. Sousa, F. Sousa, and L. Gonçalves-Seco, "Distributed database system of the New Atlas of Amphibians and Reptiles in Europe: the NA2RE project", *Amphib.-Reptil.,* vol. 35, no. 1, pp. 33-39, 2014.
[http://dx.doi.org/10.1163/15685381-00002936]

[47]   P.F. Donald, and R.J. Fuller, "Ornithological atlas data: a review of uses and limitations", *Bird Study,* vol. 2, no. 45, pp. 129-145, 1998.
[http://dx.doi.org/10.1080/00063659809461086]

[48]   P.E. Osborne, and B.J. Tigar, "Interpreting bird atlas data using logistic models: an example from Lesotho, Southern Africa", *J. Appl. Ecol.,* no. 29, pp. 55-62, 1992.
[http://dx.doi.org/10.2307/2404347]

[49]   J.M. Padial, and I. de La Riva, "Annotated checklist of the amphibians of Mauritania (West Africa)", *Rev. Esp. Herpetol.,* vol. 18, pp. 89-99, 2004.

[50]   J.M. Padial, "Commented distributional list of the reptiles of Mauritania (West Africa)", *Graellsia,* vol. 2, no. 62, pp. 159-178, 2006.
[http://dx.doi.org/10.3989/graellsia.2006.v62.i2.64]

[51]   A. Loureiro, N. Ferrand, M.A. Carretero, and O. Paulo, *Atlas dos Anfíbios e Répteis de Portugal.* Esfera do Caos: Lisboa, 2010.

[52]   J. Jalas, and J. Suonuinen, "Atlas Florae Europaea", In: *Distribution of Vascular Plants in Europe.* vol. 1. Pteridophita: Helsinki, 1972.

[53]   C. Baroni-Urbani, S. Ruffo, and A. Vigna Taglianti, "Materiali per uma biogeografia italiana fondata se alcuni generi di coleotteri cicindelidi, carabidi, e crisomelidi", *Estr Mem Soc Ent Ital,* vol. 56, pp. 35-92, 1978.

[54]   P. Sclater, "On the Geographical Distribution of the Members of the Class Aves", *J. Linn. Soc. Zool.,* no. 2, pp. 130-145, 1858.

[55]   A.R. Wallace, With a study of the relations of living and extinct faunas as elucidating the past changes of the earth's surface, *The Geographical Distribution of Animals.* Harper and Brothers: New York, 1876.

[56]   H.M. Pereira, L.M. Navarro, and I.S. Martins, "Global Biodiversity Change: The Bad, the Good, and the Unknown", *Annu. Rev. Environ. Resour.,* vol. 37, no. 1, pp. 25-50, 2012.
[http://dx.doi.org/10.1146/annurev-environ-042911-093511]

[57]   A. Galli, M. Wackernagel, K. Iha, and E. Lazarus, "Ecological footprint: Implications for biodiversity", *Biol. Conserv.,* vol. 173, pp. 121-132, 2014.
[http://dx.doi.org/10.1016/j.biocon.2013.10.019]

[58]   S.H. Butchart, M. Walpole, B. Collen, A. van Strien, J.P. Scharlemann, R.E. Almond, J.E. Baillie, B. Bomhard, C. Brown, J. Bruno, K.E. Carpenter, G.M. Carr, J. Chanson, A.M. Chenery, J. Csirke, N.C. Davidson, F. Dentener, M. Foster, A. Galli, J.N. Galloway, P. Genovesi, R.D. Gregory, M. Hockings, V. Kapos, J.F. Lamarque, F. Leverington, J. Loh, M.A. McGeoch, L. McRae, A. Minasyan, M.

Hernández Morcillo, T.E. Oldfield, D. Pauly, S. Quader, C. Revenga, J.R. Sauer, B. Skolnik, D. Spear, D. Stanwell-Smith, S.N. Stuart, A. Symes, M. Tierney, T.D. Tyrrell, J.C. Vié, and R. Watson, "Global biodiversity: indicators of recent declines", *Science,* vol. 328, no. 5982, pp. 1164-1168, 2010.
[http://dx.doi.org/10.1126/science.1187512] [PMID: 20430971]

[59]   R.A. Mittermeier, W.R. Turner, F.W. Larsen, T.M. Brooks, and C. Gascon, *Biodiversity Hotspots. Biodiversity hotspots.* Springer Berlin Heidelberg, 2011, pp. 3-22.
[http://dx.doi.org/10.1007/978-3-642-20992-5_1]

[60]   C.N. Jenkins, S.L. Pimm, and L.N. Joppa, "Global patterns of terrestrial vertebrate diversity and conservation", *Proc. Natl. Acad. Sci. USA,* vol. 110, no. 28, pp. E2602-E2610, 2013.
[http://dx.doi.org/10.1073/pnas.1302251110] [PMID: 23803854]

[61]   S. Ferrier, "Mapping spatial pattern in biodiversity for regional conservation planning: where to from here?", *Syst. Biol.,* vol. 51, no. 2, pp. 331-363, 2002.
[http://dx.doi.org/10.1080/10635150252899806] [PMID: 12028736]

[62]   H. Possingham, I. Ball, and S. Andelman, *Mathematical methods for identifying representative reserve networks.* Quant Methods Conserv Biol, 2000, pp. 291-305.
[http://dx.doi.org/10.1007/0-387-22648-6_17]

[63]   J. Grand, M.P. Cummings, T.G. Rebelo, T.H. Ricketts, and M.C. Neel, "Biased data reduce efficiency and effectiveness of conservation reserve networks", *Ecol. Lett.,* vol. 10, no. 5, pp. 364-374, 2007.
[http://dx.doi.org/10.1111/j.1461-0248.2007.01025.x] [PMID: 17498135]

[64]   N. Myers, "Threatened biotas: "hot spots" in tropical forests", *Environmentalist,* vol. 8, no. 3, pp. 187-208, 1988.
[http://dx.doi.org/10.1007/BF02240252] [PMID: 12322582]

[65]   N. Myers, "The biodiversity challenge: expanded hot-spots analysis", *Environmentalist,* vol. 10, no. 4, pp. 243-256, 1990.
[http://dx.doi.org/10.1007/BF02239720] [PMID: 12322583]

[66]   N. Myers, "Biodiversity Hotspots Revisited", *Bioscience,* vol. 53, no. 10, p. 916, 2003.
[http://dx.doi.org/10.1641/0006-3568(2003)053[0916:BHR]2.0.CO;2]

[67]   N. Myers, R.A. Mittermeier, C.G. Mittermeier, G.A. da Fonseca, and J. Kent, "Biodiversity hotspots for conservation priorities", *Nature,* vol. 403, no. 6772, pp. 853-858, 2000.
[http://dx.doi.org/10.1038/35002501] [PMID: 10706275]

[68]   D.M. Olson, E. Dinerstein, E.D. Wikramanayake, N.D. Burgess, G.V. Powell, and E.C. Underwood, "Terrestrial Ecoregions of the World: A New Map of Life on Earth", *Bioscience,* no. 51, pp. 933-938, 2001.
[http://dx.doi.org/10.1641/0006-3568(2001)051[0933:TEOTWA]2.0.CO;2]

[69]   J.M. Hoekstra, T.M. Boucher, T.H. Ricketts, and C. Roberts, "Confronting a biome crisis: Global disparities of habitat loss and protection", *Ecol. Lett.,* vol. 8, no. 1, pp. 23-29, 2005.
[http://dx.doi.org/10.1111/j.1461-0248.2004.00686.x]

[70]   A.J. Stattersfield, M.J. Crosby, A.J. Long, and D.C. Wege, *Endemic Bird Areas of the World. Priorities for Biodiversity conservation. BirdLife Conservation Series 7.* BirdLife International: Cambridge, 1998.

[71]   T.M. Brooks, R.A. Mittermeier, G.A. da Fonseca, J. Gerlach, M. Hoffmann, J.F. Lamoreux, C.G. Mittermeier, J.D. Pilgrim, and A.S. Rodrigues, "Global biodiversity conservation priorities", *Science,* vol. 313, no. 5783, pp. 58-61, 2006.
[http://dx.doi.org/10.1126/science.1127609] [PMID: 16825561]

[72]   R.A. Mittermeier, C.G. Mittermeier, T.M. Brooks, J.D. Pilgrim, W.R. Konstant, G.A. da Fonseca, and C. Kormos, "Wilderness and biodiversity conservation", *Proc. Natl. Acad. Sci. USA,* vol. 100, no. 18, pp. 10309-10313, 2003.
[http://dx.doi.org/10.1073/pnas.1732458100] [PMID: 12930898]

[73] N. Selva, S. Kreft, V. Kati, M. Schluck, B-G. Jonsson, B. Mihok, H. Okarma, and P.L. Ibisch, "Roadless and Low-Traffic Areas as Conservation Targets in Europe", *Environ. Manage.,* vol. 48, no. 5, pp. 865-877, 2011.
[http://dx.doi.org/10.1007/s00267-011-9751-z] [PMID: 21947368]

[74] E.W. Sanderson, and M. Jaiteh, "Levy M a., Redford KH, Wannebo A V., Woolmer G. The Human Footprint and the Last of the Wild", *Bioscience,* vol. 52, no. 10, p. 891, 2002.
[http://dx.doi.org/10.1641/0006-3568(2002)052[0891:THFATL]2.0.CO;2]

[75] W.V. Reid, "Biodiversity hotspots", *Trends Ecol. Evol. (Amst.),* vol. 13, no. 7, pp. 275-280, 1998.
[http://dx.doi.org/10.1016/S0169-5347(98)01363-9] [PMID: 21238297]

[76] C.G. Vale, S.L. Pimm, and J.C. Brito, "Overlooked mountain rock pools in deserts are critical local hotspots of biodiversity", *PLoS One,* vol. 10, no. 2, p. e0118367, 2015.
[http://dx.doi.org/10.1371/journal.pone.0118367] [PMID: 25714751]

[77] J.M. Scott, F. Davis, B. Csutti, R. Noss, B. Butterfield, and C. Groves, "Gap analysis: a geographic approach to protection of biological diversity", *Wildl. Monogr.,* no. 123, pp. 1-41, 1993.

[78] J.M. Scott, F.W. Davis, R.G. McGhie, R.G. Wright, C. Groves, and J. Estes, "Nature reserves: Do they capture the full range of America's biological diversity?", *Ecol. Appl.,* vol. 11, no. 4, pp. 999-1007, 2001.
[http://dx.doi.org/10.1890/1051-0761(2001)011[0999:NRDTCT]2.0.CO;2]

[79] S.J. Andelman, and M.R. Willig, "Present patterns and future prospects for biodiversity in the Western Hemisphere", *Ecol. Lett.,* vol. 6, no. 9, pp. 818-824, 2003.
[http://dx.doi.org/10.1046/j.1461-0248.2003.00503.x]

[80] A.S. Rodrigues, S.J. Andelman, M.I. Bakarr, L. Boitani, T.M. Brooks, R.M. Cowling, L.D. Fishpool, G.A. Da Fonseca, K.J. Gaston, M. Hoffmann, J.S. Long, P.A. Marquet, J.D. Pilgrim, R.L. Pressey, J. Schipper, W. Sechrest, S.N. Stuart, L.G. Underhill, R.W. Waller, M.E. Watts, and X. Yan, "Effectiveness of the global protected area network in representing species diversity", *Nature,* vol. 428, no. 6983, pp. 640-643, 2004.
[http://dx.doi.org/10.1038/nature02422] [PMID: 15071592]

[81] P. Hugh, "Possingham, Wilson KA. Turning up the heat on myeloma", *Nature,* vol. 436, pp. 919-920, 2005.
[PMID: 16107821]

[82] K.A. Wilson, M.F. McBride, M. Bode, and H.P. Possingham, "Prioritizing global conservation efforts", *Nature,* vol. 440, no. 7082, pp. 337-340, 2006.
[http://dx.doi.org/10.1038/nature04366] [PMID: 16541073]

[83] C.R. Margules, and R.L. Pressey, "Systematic conservation planning", *Nature,* vol. 405, no. 6783, pp. 243-253, 2000.
[http://dx.doi.org/10.1038/35012251] [PMID: 10821285]

[84] A. Moilanen, and H. Kujala, The Zonation framework and software for conservation prioritization.*Spatial conservation prioritization: Quantitative Methods and Computational Tools* Oxford University Press, 2009, pp. 196-210.A. Moilanen, K.A. Wilson, and H.P. Possingham,

[85] S.A. Smith, P.R. Stephens, and J.J. Wiens, "Replicate patterns of species richness, historical biogeography, and phylogeny in Holarctic treefrogs", *Evolution,* vol. 59, no. 11, pp. 2433-2450, 2005.
[http://dx.doi.org/10.1111/j.0014-3820.2005.tb00953.x] [PMID: 16396184]

[86] R.M. Cowling, R.L. Pressey, A.T. Lombard, P.G. Desmet, and A.G. Ellis, "From representation to persistence: Requirements for a sustainable system of conservation areas in the species-rich Mediterranean-climate desert of southern Africa", *Divers. Distrib.,* vol. 5, no. 1-2, pp. 51-71, 1999.
[http://dx.doi.org/10.1046/j.1472-4642.1999.00038.x]

[87] TM Brooks, MI Bakarr, and T Boucher, "Coverage Provided by the Global Protected-Area System: Is It Enough?", *Bioscience,* vol. 54, no. 12, p. 1081, 2004.

[http://dx.doi.org/10.1641/0006-3568(2004)054[1081:CPBTGP]2.0.CO;2]

[88]   S. Ferrier, G. Watson, J. Pearce, and M. Drielsma, "Extended statistical approaches to modelling spatial pattern in biodiversity in northeast New SouthWales. I. Species-level modelling", *Biodivers. Conserv.*, no. 11, pp. 2275-2307, 2002.
[http://dx.doi.org/10.1023/A:1021302930424]

[89]   M. Cabeza, and A. Moilanen, "Site-Selection Algorithms and Habitat Loss", *Conserv. Biol.*, vol. 17, no. 5, pp. 1402-1413, 2003.
[http://dx.doi.org/10.1046/j.1523-1739.2003.01421.x]

[90]   S. Sarkar, R. Pressey, D. Faith, C. Margules, T. Fuller, and D. Stoms, "Biodiversity Conservation Planning Tools: Present Status and Challenges for the Future", *Annu. Rev. Environ. Resour.*, vol. 31, no. 1, pp. 123-159, 2006.
[http://dx.doi.org/10.1146/annurev.energy.31.042606.085844]

[91]   J.C. Williams, C.S. ReVelle, and S.A. Levin, "Using mathematical optimization models to design nature reserves", *Front Ecol. Environ.*, vol. 2, no. 2Nr: 72, pp. 98-105, 2004.

[92]   J.B. Kirkpatrick, "An iterative method for establishing priorities for the selection of nature reserves: An example from Tasmania", *Biol. Conserv.*, vol. 25, no. 2, pp. 127-134, 1983.
[http://dx.doi.org/10.1016/0006-3207(83)90056-3]

[93]   R.L. Pressey, "Classics in physical geography revisited", *Prog. Phys. Geogr.*, vol. 26, no. 3, pp. 434-441, 2002.
[http://dx.doi.org/10.1191/0309133302pp347xx]

[94]   R.L. Pressey, C.J. Humphries, C.R. Margules, R.I. Vane-Wright, and P.H. Williams, "Beyond opportunism: Key principles for systematic reserve selection", *Trends Ecol. Evol. (Amst.)*, vol. 8, no. 4, pp. 124-128, 1993.
[http://dx.doi.org/10.1016/0169-5347(93)90023-I] [PMID: 21236127]

[95]   J. Arthur, M. Hackey, K. Sahr, M. Huso, and A. Kiester, "Finding all optimal solutions to the reserve site selection problem: formulation and computational analysis", *Environ. Ecol. Stat.*, vol. 4, p. 153, 1997.
[http://dx.doi.org/10.1023/A:1018570311399]

[96]   R.L. Church, D.M. Stoms, and F.W. Davis, "Reserve selection as a maximal covering location problem", *Biol. Conserv.*, vol. 76, no. 2, pp. 105-112, 1996.
[http://dx.doi.org/10.1016/0006-3207(95)00102-6]

[97]   J.D. Camm, S. Polasky, A. Solow, and B. Csuti, "A note on optimal algorithms for reserve site selection", *Biol. Conserv.*, vol. 78, no. 3, pp. 353-355, 1996.
[http://dx.doi.org/10.1016/0006-3207(95)00132-8]

[98]   I.R. Ball, H.P. Possingham, and M.E. Watts, "Marxan and Relatives: Software for Spatial Conservation Prioritization", *Spat. Conserv. Prioritization Quant. Methods Comput. Tools*, pp. 185-195, 2009.

[99]   S.B. Carvalho, J.C. Brito, E.G. Crespo, M.E. Watts, and H.P. Possingham, "Conservation planning under climate change: Toward accounting for uncertainty in predicted species distributions to increase confidence in conservation investments in space and time", *Biol. Conserv.*, vol. 144, no. 7, pp. 2020-2030, 2011.
[http://dx.doi.org/10.1016/j.biocon.2011.04.024]

[100] J.C. Brito, P. Tarroso, C.G. Vale, F. Martínez-Freiría, Z. Boratyński, and J.C. Campos, "Conservation Biogeography of the Sahara-Sahel: Additional protected areas are needed to secure unique biodiversity", *Divers. Distrib.*, pp. 1-14, 2016.

[101] A. Arponen, R.K. Heikkinen, R. Paloniemi, J. Pöyry, J. Similä, and M. Kuussaari, "Improving conservation planning for semi-natural grasslands: Integrating connectivity into agri-environment schemes", *Biol. Conserv.*, vol. 160, pp. 234-241, 2013.

[http://dx.doi.org/10.1016/j.biocon.2013.01.018]

[102] E. Di Minin, D.C. Macmillan, P.S. Goodman, B. Escott, R. Slotow, and A. Moilanen, "Conservation businesses and conservation planning in a biological diversity hotspot", *Conserv. Biol.,* vol. 27, no. 4, pp. 808-820, 2013.
[http://dx.doi.org/10.1111/cobi.12048] [PMID: 23565917]

[103] J. Elith, and J.R. Leathwick, "Species Distribution Models: Ecological Explanation and Prediction Across Space and Time", *Annu. Rev. Ecol. Evol. Syst.,* vol. 40, no. 1, pp. 677-697, 2009.
[http://dx.doi.org/10.1146/annurev.ecolsys.110308.120159]

[104] J. Franklin, *Mapping Species Distributions.* Cambridge University Press, 2010.
[http://dx.doi.org/10.1017/CBO9780511810602]

[105] A. Guisan, R. Tingley, J.B. Baumgartner, I. Naujokaitis-Lewis, P.R. Sutcliffe, A.I. Tulloch, T.J. Regan, L. Brotons, E. McDonald-Madden, C. Mantyka-Pringle, T.G. Martin, J.R. Rhodes, R. Maggini, S.A. Setterfield, J. Elith, M.W. Schwartz, B.A. Wintle, O. Broennimann, M. Austin, S. Ferrier, M.R. Kearney, H.P. Possingham, and Y.M. Buckley, "Predicting species distributions for conservation decisions", *Ecol. Lett.,* vol. 16, no. 12, pp. 1424-1435, 2013.
[http://dx.doi.org/10.1111/ele.12189] [PMID: 24134332]

[106] J. Elith, C. Graham, R. Anderson, M. Dudik, S. Ferrier, and A. Guisan, "Novel methods improve prediction of species' distributions from occurrence data", *Ecography,* vol. 29, no. 2, pp. 129-151, 2006.
[http://dx.doi.org/10.1111/j.2006.0906-7590.04596.x]

[107] M.B. Araujo, and A. Guisan, "Five (or so) challenges for species distribution modelling", *J. Biogeogr.,* vol. 10, no. 33, pp. 1677-1688, 2006.
[http://dx.doi.org/10.1111/j.1365-2699.2006.01584.x]

[108] J.A. Wiens, D. Stralberg, D. Jongsomjit, C.A. Howell, and M.A. Snyder, "Niches, models, and climate change: assessing the assumptions and uncertainties", *Proc. Natl. Acad. Sci. USA,* vol. 106, suppl. Suppl. 2, pp. 19729-19736, 2009.
[http://dx.doi.org/10.1073/pnas.0901639106] [PMID: 19822750]

[109] C.B. Yackulic, R. Chandler, E.F. Zipkin, J.A. Royle, D. James, and E.H. Grant, "Presence-only modelling using MAXENT: when can we trust the inferences?", *Methods Ecol. Evol.,* vol. 4, pp. 236-243, 2013.
[http://dx.doi.org/10.1111/2041-210x.12004]

[110] C.G. Vale, P. Tarroso, and J.C. Brito, "Predicting species distribution at range margins: Testing the effects of study area extent, resolution and threshold selection in the Sahara-Sahel transition zone", *Divers. Distrib.,* vol. 20, no. 1, pp. 20-33, 2014.
[http://dx.doi.org/10.1111/ddi.12115]

[111] C.G. Vale, M.J. Ferreira da Silva, J.C. Campos, J. Torres, and J.C. Brito, "Applying species distribution modelling to the conservation of an ecologically plastic species (Papio papio) across biogeographic regions in West Africa", *J. Nat. Conserv.,* vol. 27, pp. 26-36, 2015.
[http://dx.doi.org/10.1016/j.jnc.2015.06.004]

[112] P. Gaubert, M. Papes, and A.T. Peterson, "Natural history collections and the conservation of poorly known taxa: Ecological niche modeling in central African rainforest genets (Genetta spp.)", *Biol. Conserv.,* vol. 1, no. 130, pp. 106-117, 2006.
[http://dx.doi.org/10.1016/j.biocon.2005.12.006]

[113] A. Travaini, M. Delibes, and P. Ferreras, "Diversity, abundance or rare species as a target for the conservation of mammalian carnivores: a case study in Southern Spain", *Biodivers. Conserv.,* vol. 535, pp. 529-535, 1997.
[http://dx.doi.org/10.1023/A:1018329127772]

[114] C.G. Vale, P. Tarroso, J.C. Campos, D. Vasconcelos Gonçalves, and J.C. Brito, "Distribution, suitable areas and conservation status of the Boulenger's agama (Agama boulengeri, Lataste 1886)",

Amphibia-Reptilia, vol. 33, no. 2, pp. 526-532, 2012.
[http://dx.doi.org/10.1163/15685381-00002853]

[115] C.G. Vale, F. Álvares, and J.C. Brito, "Distribution, suitable areas and conservation status of the Felou gundi (Felovia vae Lataste 1886)", *Mammalia,* vol. 76, no. 2, pp. 201-207, 2012.
[http://dx.doi.org/10.1515/mammalia-2011-0008]

[116] F. Martínez-Freiría, P. Tarroso, H. Rebelo, and J.C. Brito, "Contemporary niche contraction affects climate change predictions for elephants and giraffes", *Divers. Distrib.,* pp. 1-13, 2015.

[117] J.R. Leathwick, and M.P. Austin, "Competitive interactions between tree species in New Zealand old-growth indigenous forests", *Ecology,* vol. 82, no. 9, pp. 2560-2573, 2001.
[http://dx.doi.org/10.1890/0012-9658(2001)082[2560:CIBTSI]2.0.CO;2]

[118] C.H. Graham, S.R. Ron, J.C. Santos, C.J. Schneider, and C. Moritz, "Integrating phylogenetics and environmental niche models to explore speciation mechanisms in dendrobatid frogs", *Evolution,* vol. 58, no. 8, pp. 1781-1793, 2004.
[http://dx.doi.org/10.1111/j.0014-3820.2004.tb00461.x] [PMID: 15446430]

[119] J.C. Habel, F.E. Zachos, L. Dapporto, D. Rodder, U. Radespiel, and A. Tellier, "Population genetics revisited - towards a multidisciplinary research field", *Biol. J. Linn. Soc. Lond.,* vol. 115, no. 1, pp. 1-12, 2015.
[http://dx.doi.org/10.1111/bij.12481]

[120] R. Vodă, L. Dapporto, V. Dincă, and R. Vila, "Why do cryptic species tend not to co-occur? A case study on two cryptic pairs of butterflies", *PLoS One,* vol. 10, no. 2, p. e0117802, 2015.
[http://dx.doi.org/10.1371/journal.pone.0117802] [PMID: 25692577]

[121] R.E. Glor, and D. Warren, "Testing ecological explanations for biogeographic boundaries", *Evolution,* vol. 65, no. 3, pp. 673-683, 2011.
[http://dx.doi.org/10.1111/j.1558-5646.2010.01177.x] [PMID: 21054358]

[122] M.B. Araujo, W. Thuiller, and R.G. Pearson, "Climate warming and the decline of amphibians and reptiles in Europe", *J. Biogeogr.,* no. 33, pp. 1712-1728, 2006.
[http://dx.doi.org/10.1111/j.1365-2699.2006.01482.x]

[123] P.F. Addison, L. Rumpff, S.S. Bau, J.M. Carey, Y.E. Chee, and F.C. Jarrad, "Practical solutions for making models indispensable in conservation decision-making", *Divers. Distrib.,* vol. 19, no. 5-6, pp. 490-502, 2013.
[http://dx.doi.org/10.1111/ddi.12054]

[124] C.M. Beale, J.J. Lennon, and A. Gimona, "Opening the climate envelope reveals no macroscale associations with climate in European birds", *Proc. Natl. Acad. Sci. USA,* vol. 105, no. 39, pp. 14908-14912, 2008.
[http://dx.doi.org/10.1073/pnas.0803506105] [PMID: 18815364]

[125] CM Beale, and JJ Lennon, "Incorporating uncertainty in predictive species distribution modelling", *Philos. Trans. R. Soc. Lond. B Biol. Sci.,* vol. 367, no. 1586, pp. 247-258, 1586.

[126] A.T. Peterson, J. Soberón, and R.G. Pearson, *Ecological Niche and Geographical Distributions.* Princenton University Press: New Jersey, 2011.

[127] P.V. Wells, and C.D. Jorgensen, "Pleistocene Wood Rat Middens and Climatic Change in Mohave Desert: A Record of Juniper Woodlands", *Science,* vol. 143, no. 3611, pp. 1171-1173, 1964.
[http://dx.doi.org/10.1126/science.143.3611.1171] [PMID: 17833902]

[128] G.R. Walther, S. Berger, and M.T. Sykes, "An ecological "footprint" of climate change", *Proc. R. Soc. B. Biol. Sci.,* vol. 272, no. 1571, pp. 1427-1432, 2005.

[129] C.D. Thomas, and J.J. Lennon, "Birds extend their ranges northwards", *Nature,* vol. 399, p. 213, 1999.
[http://dx.doi.org/10.1038/20335]

[130] C. Parmesan, "Ecological and Evolutionary Responses to Recent Climate Change", *Annu. Rev. Ecol.*

Evol. Syst., vol. 1, no. 37, pp. 637-669, 2006.
[http://dx.doi.org/10.1146/annurev.ecolsys.37.091305.110100]

[131] BFN Erasmus, AS Van Jaarsveld, SL Chown, M Kshatriya, and KJ Wessels, "Vulnerability of South African animal taxa to climate change", *Glob. Chang. Biol.,* no. 8, pp. 679-693, 2002.
[http://dx.doi.org/10.1046/j.1365-2486.2002.00502.x]

[132] A.T. Peterson, M.A. Ortega-Huerta, J. Bartley, V. Sánchez-Cordero, J. Soberón, R.H. Buddemeier, and D.R. Stockwell, "Future projections for Mexican faunas under global climate change scenarios", *Nature,* vol. 416, no. 6881, pp. 626-629, 2002.
[http://dx.doi.org/10.1038/416626a] [PMID: 11948349]

[133] W. Thuiller, S. Lavorel, M.B. Araújo, M.T. Sykes, and I.C. Prentice, "Climate change threats to plant diversity in Europe", *Proc. Natl. Acad. Sci. USA,* vol. 102, no. 23, pp. 8245-8250, 2005.
[http://dx.doi.org/10.1073/pnas.0409902102] [PMID: 15919825]

[134] I-C. Chen, J.K. Hill, R. Ohlemüller, D.B. Roy, and C.D. Thomas, "Rapid range shifts of species associated with high levels of climate warming", *Science,* vol. 333, no. 6045, pp. 1024-1026, 2011.
[http://dx.doi.org/10.1126/science.1206432] [PMID: 21852500]

[135] W.B. Foden, S.H. Butchart, S.N. Stuart, J.C. Vié, H.R. Akçakaya, A. Angulo, L.M. DeVantier, A. Gutsche, E. Turak, L. Cao, S.D. Donner, V. Katariya, R. Bernard, R.A. Holland, A.F. Hughes, S.E. O'Hanlon, S.T. Garnett, C.H. Sekercioğlu, and G.M. Mace, "Identifying the world's most climate change vulnerable species: a systematic trait-based assessment of all birds, amphibians and corals", *PLoS One,* vol. 8, no. 6, p. e65427, 2013.
[http://dx.doi.org/10.1371/journal.pone.0065427] [PMID: 23950785]

[136] C.G. Vale, and J.C. Brito, "Desert-adapted species are vulnerable to climate change: Insights from the warmest region on Earth", *Glob. Ecol. Conserv.,* vol. 4, pp. 369-379, 2015.
[http://dx.doi.org/10.1016/j.gecco.2015.07.012]

[137] J.J. Lawler, S.L. Shafer, B.A. Bancroft, and A.R. Blaustein, "Projected climate impacts for the amphibians of the Western hemisphere", *Conserv. Biol.,* vol. 24, no. 1, pp. 38-50, 2010.
[http://dx.doi.org/10.1111/j.1523-1739.2009.01403.x] [PMID: 20121840]

[138] S.R. Loarie, P.B. Duffy, H. Hamilton, G.P. Asner, C.B. Field, and D.D. Ackerly, "The velocity of climate change", *Nature,* vol. 462, no. 7276, pp. 1052-1055, 2009.
[http://dx.doi.org/10.1038/nature08649] [PMID: 20033047]

[139] M. Parida, A.A. Hoffmann, and M.P. Hill, "Climate change expected to drive habitat loss for two key herbivore species in an alpine environment", *J. Biogeogr.,* pp. 1210-1221, 2015.
[http://dx.doi.org/10.1111/jbi.12490]

[140] D.A. Prieto-Torres, A.G. Navarro-Sigüenza, D. Santiago-Alarcon, and O.R. Rojas-Soto, *Response of the endangered tropical dry forests to climate change and the role of Mexican Protected Areas for their conservation.* Glob Chang Biol, 2015, pp. 364-379.

[141] M.B. Araújo, M. Cabeza, W. Thuiller, L. Hannah, and P.H. Williams, "Would climate change drive species out of reserves? An assessment of existing reserve-selection methods", *Glob. Change Biol.,* vol. 10, no. 9, pp. 1618-1626, 2004.
[http://dx.doi.org/10.1111/j.1365-2486.2004.00828.x]

[142] J. Soberón, and A.T. Peterson, "Interpretation of Models of Fundamental Ecological Niches and Species Distributional Areas", *Biodivers. Inform.,* vol. 2005, no. 2, pp. 1-10, 2005.

[143] R.K. Meentemeyer, B.L. Anacker, W. Mark, and D.M. Rizzo, "Early detection of emerging forest disease using dispersal estimation and ecological niche modeling", *Ecol. Appl.,* vol. 18, no. 2, pp. 377-390, 2008.
[http://dx.doi.org/10.1890/07-1150.1] [PMID: 18488603]

[144] N. Barve, V. Barve, A. Jiménez-Valverde, A. Lira-Noriega, S.P. Maher, and T. Peterson, "The crucial role of the accessible area in ecological niche modeling and species distribution modeling", *Ecol.*

Modell., vol. 222, no. 11, pp. 1810-1819, 2011.
[http://dx.doi.org/10.1016/j.ecolmodel.2011.02.011]

[145] T. Vaclavik, and R.K. Meentemeyer, "Invasive species distribution modeling (iSDM): Are absence data and dispersal constraints needed to predict actual distributions?", *Ecol. Modell.,* vol. 220, no. 23, pp. 3248-3258, 2009.
[http://dx.doi.org/10.1016/j.ecolmodel.2009.08.013]

[146] L.R. Iverson, A.M. Prasad, S.N. Matthews, and M.P. Peters, "Lessons Learned While Integrating Habitat, Dispersal, Disturbance, and Life-History Traits into Species Habitat Models Under Climate Change", *Ecosystems (N. Y.),* vol. 14, no. 6, pp. 1005-1020, 2011.
[http://dx.doi.org/10.1007/s10021-011-9456-4]

[147] P. de Pous, A. Montori, F. Amat, and D. Sanuy, *Range contraction and loss of genetic variation of the Pyrenean endemic newt Calotriton asper due to climate change.* Reg Environ Chang, 2015.

[148] G.F. Midgley, I.D. Davies, C.H. Albert, R. Altwegg, L. Hannah, and G.O. Hughes, "BioMove - an integrated platform simulating the dynamic response of species to environmental change", *Ecography,* vol. 33, no. 3, pp. 612-616, 2010.

[149] R. Engler, W. Hordijk, and A. Guisan, "The MIGCLIM R package - seamless integration of dispersal constraints into projections of species distribution models", *Ecography,* vol. 35, no. 10, pp. 872-878, 2012.
[http://dx.doi.org/10.1111/j.1600-0587.2012.07608.x]

[150] D.F. Sax, J.J. Stachowicz, J.H. Brown, J.F. Bruno, M.N. Dawson, S.D. Gaines, R.K. Grosberg, A. Hastings, R.D. Holt, M.M. Mayfield, M.I. O'Connor, and W.R. Rice, "Ecological and evolutionary insights from species invasions", *Trends Ecol. Evol. (Amst.),* vol. 22, no. 9, pp. 465-471, 2007.
[http://dx.doi.org/10.1016/j.tree.2007.06.009] [PMID: 17640765]

[151] N. Roura-Pascual, A.V. Suárez, C. Gómez, P. Pons, Y. Touyama, A.L. Wild, and A.T. Peterson, "Geographical potential of Argentine ants (Linepithema humile Mayr) in the face of global climate change", *Proc. Biol. Sci.,* vol. 271, no. 1557, pp. 2527-2535, 2004.
[http://dx.doi.org/10.1098/rspb.2004.2898] [PMID: 15615677]

[152] A.T. Peterson, and D.A. Vieglais, "Predicting Species Invasions Using Ecological Niche Modeling: New Approaches from Bioinformatics Attack a Pressing Problem", *Bioscience,* vol. 5, no. 51, pp. 363-371, 2001.
[http://dx.doi.org/10.1641/0006-3568(2001)051[0363:PSIUEN]2.0.CO;2]

[153] R. Real, A.L. Marquez, A. Estrada, A.R. Munoz, and J.M. Vargas, "Modelling chorotypes of invasive vertebrates in mainland Spain", *Divers. Distrib.,* vol. 2, no. 14, pp. 364-373, 2008.

[154] N.G. Swenson, "The past and future influence of geographic information systems on hybrid zone, phylogeographic and speciation research", *J. Evol. Biol.,* vol. 21, no. 2, pp. 421-434, 2008.
[http://dx.doi.org/10.1111/j.1420-9101.2007.01487.x] [PMID: 18205783]

[155] B. Kohlmann, H. Nix, and D.D. Shaw, "Environmental predictions and distributional limits of chromosomal taxa in the Australian grasshopperCaledia captiva (F.)", *Oecologia,* vol. 75, no. 4, pp. 483-493, 1988.
[http://dx.doi.org/10.1007/BF00776409] [PMID: 28312420]

[156] M.W. Chatfield, K.H. Kozak, B.M. Fitzpatrick, and P.K. Tucker, "Patterns of differential introgression in a salamander hybrid zone: inferences from genetic data and ecological niche modelling", *Mol. Ecol.,* vol. 19, no. 19, pp. 4265-4282, 2010.
[http://dx.doi.org/10.1111/j.1365-294X.2010.04796.x] [PMID: 20819165]

[157] Z.W. Culumber, D.B. Shepard, S.W. Coleman, G.G. Rosenthal, and M. Tobler, "Physiological adaptation along environmental gradients and replicated hybrid zone structure in swordtails (Teleostei: Xiphophorus)", *J. Evol. Biol.,* vol. 25, no. 9, pp. 1800-1814, 2012.
[http://dx.doi.org/10.1111/j.1420-9101.2012.02562.x] [PMID: 22827312]

[158] M.E. Blair, E.J. Sterling, M. Dusch, C.J. Raxworthy, and R.G. Pearson, "Ecological divergence and speciation between lemur (Eulemur) sister species in Madagascar", *J. Evol. Biol.,* vol. 26, no. 8, pp. 1790-1801, 2013.
[http://dx.doi.org/10.1111/jeb.12179] [PMID: 23865477]

[159] S.E. Johnson, K.E. Delmore, K.A. Brown, T.M. Wyman, and E.E. Louis, "Niche Divergence in a Brown Lemur (Eulemur spp.) Hybrid Zone: Using Ecological Niche Models to Test Models of Stability", *Int. J. Primatol.,* vol. 73, no. 1, pp. 69-88, 2016.
[http://dx.doi.org/10.1007/s10764-015-9872-y]

[160] P. Tarroso, R.J. Pereira, F. Martínez-Freiría, R. Godinho, and J.C. Brito, "Hybridization at an ecotone: ecological and genetic barriers between three Iberian vipers", *Mol. Ecol.,* vol. 23, no. 5, pp. 1108-1123, 2014.
[http://dx.doi.org/10.1111/mec.12671] [PMID: 24447270]

[161] K.H. Kozak, C.H. Graham, and J.J. Wiens, "Integrating GIS-based environmental data into evolutionary biology", *Trends Ecol. Evol. (Amst.),* vol. 23, no. 3, pp. 141-148, 2008.
[http://dx.doi.org/10.1016/j.tree.2008.02.001] [PMID: 18291557]

[162] D.F. Alvarado-Serrano, and L.L. Knowles, "Ecological niche models in phylogeographic studies: applications, advances and precautions", *Mol. Ecol. Resour.,* vol. 14, no. 2, pp. 233-248, 2014.
[http://dx.doi.org/10.1111/1755-0998.12184] [PMID: 24119244]

[163] R.J. Hijmans, E. Cameron, J.L. Parra, P.G. Jones, and A. Jarvis, "Very high resolution interpolated climate surfaces for global land areas", *Int. J. Climatol.,* vol. 25, no. 15, pp. 1965-1978, 2005.
[http://dx.doi.org/10.1002/joc.1276]

[164] E. Waltari, R.J. Hijmans, A.T. Peterson, Á.S. Nyári, S.L. Perkins, and R.P. Guralnick, "Locating pleistocene refugia: comparing phylogeographic and ecological niche model predictions", *PLoS One,* vol. 2, no. 6, p. e563, 2007.
[http://dx.doi.org/10.1371/journal.pone.0000563] [PMID: 17622339]

[165] B.C. Carstens, and C.L. Richards, *Integrating Coalescent and Ecological Niche Modeling in Comparative Phylogeography,* vol. 61, no. 6, pp. 1439-1454, 2007.

[166] L.L. Knowles, and D.F. Alvarado-Serrano, "Exploring the population genetic consequences of the colonization process with spatio-temporally explicit models: insights from coupled ecological, demographic and genetic models in montane grasshoppers", *Mol. Ecol.,* vol. 19, no. 17, pp. 3727-3745, 2010.
[http://dx.doi.org/10.1111/j.1365-294X.2010.04702.x] [PMID: 20723059]

[167] J.A. Banta, I.M. Ehrenreich, S. Gerard, L. Chou, A. Wilczek, J. Schmitt, P.X. Kover, and M.D. Purugganan, "Climate envelope modelling reveals intraspecific relationships among flowering phenology, niche breadth and potential range size in Arabidopsis thaliana", *Ecol. Lett.,* vol. 15, no. 8, pp. 769-777, 2012.
[http://dx.doi.org/10.1111/j.1461-0248.2012.01796.x] [PMID: 22583905]

[168] D.L. Edwards, J.S. Keogh, and L.L. Knowles, "Effects of vicariant barriers, habitat stability, population isolation and environmental features on species divergence in the south-western Australian coastal reptile community", *Mol. Ecol.,* vol. 21, no. 15, pp. 3809-3822, 2012.
[http://dx.doi.org/10.1111/j.1365-294X.2012.05637.x] [PMID: 22646317]

[169] K. De Queiroz, "Species concepts and species delimitation", *Syst. Biol.,* vol. 56, no. 6, pp. 879-886, 2007.
[http://dx.doi.org/10.1080/10635150701701083] [PMID: 18027281]

[170] D.L. Warren, R.E. Glor, and M. Turelli, "Environmental niche equivalency versus conservatism: quantitative approaches to niche evolution", *Evolution,* vol. 62, no. 11, pp. 2868-2883, 2008.
[http://dx.doi.org/10.1111/j.1558-5646.2008.00482.x] [PMID: 18752605]

[171] O. Broennimann, M.C. Fitzpatrick, P.B. Pearman, B. Petitpierre, L. Pellissier, and N.G. Yoccoz,

"Measuring ecological niche overlap from occurrence and spatial environmental data", *Glob. Ecol. Biogeogr.,* vol. 21, no. 4, pp. 481-497, 2012.
[http://dx.doi.org/10.1111/j.1466-8238.2011.00698.x]

[172] J.W. Fitzpatrick, "Subspecies are for Convenience", *Ornithol. Monogr.,* vol. 67, no. 1, pp. 54-61, 2010.
[http://dx.doi.org/10.1525/om.2010.67.1.54]

[173] L.J. Rissler, and J.J. Apodaca, "Adding more ecology into species delimitation: ecological niche models and phylogeography help define cryptic species in the black salamander (Aneides flavipunctatus)", *Syst. Biol.,* vol. 56, no. 6, pp. 924-942, 2007.
[http://dx.doi.org/10.1080/10635150701703063] [PMID: 18066928]

[174] B. Wielstra, W. Beukema, J.W. Arntzen, A.K. Skidmore, A.G. Toxopeus, and N. Raes, "Corresponding mitochondrial DNA and niche divergence for crested newt candidate species", *PLoS One,* vol. 7, no. 9, p. e46671, 2012.
[http://dx.doi.org/10.1371/journal.pone.0046671] [PMID: 23029564]

[175] T.A. Pelletier, C. Crisafulli, S. Wagner, A.J. Zellmer, and B.C. Carstens, "Historical species distribution models predict species limits in western plethodon salamanders", *Syst. Biol.,* vol. 64, no. 6, pp. 909-925, 2015.
[http://dx.doi.org/10.1093/sysbio/syu090] [PMID: 25414176]

[176] M.F. Castillo-Cardenas, J.A. Ramirez-Silva, O. Sanjur, and N. Toro-Perea, "Evidence of incipient speciation in the Neotropical mangrove Pelliciera rhizophorae (Tetrameristaceae) as revealed by molecular, morphological, physiological and climatic characteristics", *Bot. J. Linn. Soc.,* vol. 179, pp. 499-510, 2015.
[http://dx.doi.org/10.1111/boj.12337]

[177] L. Fahrig, "Effects of habitat fragmentation on biodiversity", *Annu. Rev. Ecol. Evol. Syst.,* vol. 34, pp. 487-515, 2003.
[http://dx.doi.org/10.1146/annurev.ecolsys.34.011802.132419]

[178] J. Fischer, and D.B. Lindenmayer, "Landscape modification and habitat fragmentation: a synthesis", *Glob. Ecol. Biogeogr.,* vol. 16, pp. 265-280, 2007.
[http://dx.doi.org/10.1111/j.1466-8238.2007.00287.x]

[179] R. Ribeiro, X. Santos, N. Sillero, M.A. Carretero, and G.A. Llorente, "Biodiversity and Land Uses: Is Agriculture the Biggest Threat for reptiles' assemblages?", *Acta Oecol.,* vol. 35, pp. 327-334, 2009.
[http://dx.doi.org/10.1016/j.actao.2008.12.003]

[180] E.S. Minor, and D.L. Urban, "A graph-theory framework for evaluating landscape connectivity and conservation planning", *Conserv. Biol.,* vol. 22, no. 2, pp. 297-307, 2008.
[http://dx.doi.org/10.1111/j.1523-1739.2007.00871.x] [PMID: 18241238]

[181] M.D. Cantwell, and R.T. Forman, "Landscape graphs: Ecological modeling with graph theory to detect configurations common to diverse landscapes", *Landsc. Ecol.,* vol. 4, no. 8, pp. 239-255, 1993.
[http://dx.doi.org/10.1007/BF00125131]

[182] D. Urban, and T. Keitt, "Landscape Connectivity: A Graph-Theoretic Perspective", *Ecology,* vol. 5, no. 82, pp. 1205-1218, 2001.
[http://dx.doi.org/10.1890/0012-9658(2001)082[1205:LCAGTP]2.0.CO;2]

[183] N. Ray, A. Lehmann, and P. Joly, "Modeling spatial distribution of amphibian populations: a GIS approach based on habitat matrix permeability", *Biodivers. Conserv.,* no. 11, pp. 2143-2164, 2002.
[http://dx.doi.org/10.1023/A:1021390527698]

[184] K. Osterwalder, A. Klingenbock, and R. Shine, "Field studies on a social lizard: Home range and social organization in an Australian skink, Egernia major", *Austral Ecol.,* vol. 29, no. 3, pp. 241-249, 2004.
[http://dx.doi.org/10.1111/j.1442-9993.2004.01339.x]

[185] W.H. Burt, "Territoriality and Home Range Concepts as Applied to Mammals", *J. Mammal.,* vol. 24, no. 3, pp. 346-352, 1943.
[http://dx.doi.org/10.2307/1374834]

[186] P.N. Laver, and M.J. Kelly, "A Critical Review of Home Range Studies", *J. Wildl. Manage.,* vol. 72, no. 1, pp. 290-298, 2008.
[http://dx.doi.org/10.2193/2005-589]

[187] M.L. Reid, and P.J. Weatherhead, "Topographical Constraints on Competition for Territories", *Oikos,* vol. 1, no. 51, pp. 115-117, 1988.
[http://dx.doi.org/10.2307/3565819]

[188] P.A. Powell, and M.S. Mitchell, "Topographical constraints and home range quality", *Ecography,* vol. 4, no. 21, pp. 337-341, 1998.
[http://dx.doi.org/10.1111/j.1600-0587.1998.tb00398.x]

[189] P. Monterroso, N. Sillero, L.M. Rosalino, F. Loureiro, and P.C. Alves, "Estimating home-range size: when to include a third dimension?", *Ecol. Evol.,* vol. 3, no. 7, pp. 2285-2295, 2013.
[http://dx.doi.org/10.1002/ece3.590] [PMID: 23919170]

[190] S.W. Selkirk, and I.D. Bishop, "Improving and Extending Home Range and Habitat Analysis by Integration with a Geographic Information System", *Trans. GIS,* vol. 6, no. 2, pp. 151-159, 2002.
[http://dx.doi.org/10.1111/1467-9671.00102]

[191] W.E. Cooper Jr, "Home range size and population dynamics", *J. Theor. Biol.,* vol. 75, no. 3, pp. 327-337, 1978.
[http://dx.doi.org/10.1016/0022-5193(78)90338-7] [PMID: 745446]

[192] M. Warren, M.P. Robertson, and J.M. Greeff, "A comparative approach to understanding factors limiting abundance patterns and distributions in a fig tree-fig wasp mutualism", *Ecography,* vol. 1, no. 33, pp. 148-158, 2010.
[http://dx.doi.org/10.1111/j.1600-0587.2009.06041.x]

[193] E.R. Pianka, "The structure of lizard communities", *Annu. Rev. Ecol. Syst.,* vol. 4, pp. 53-74, 1973.
[http://dx.doi.org/10.1146/annurev.es.04.110173.000413]

[194] R.E. Gorton, J. Fulmer, and W.J. Bell, "Spacing Patterns and Dominance in the Cockroach, Eublaberus posticus (Dictyoptera: Blaberidae)", *J. Kans. Entomol. Soc.,* vol. 52, no. 2, pp. 334-343, 1979.

[195] C.L. Frost, and P.J. Bergmann, "Spatial Distribution and Habitat Utilization of the Zebra-tailed Lizard (Callisaurus draconoides)", *J. Herpetol.,* vol. 46, no. 2, pp. 203-208, 2012.
[http://dx.doi.org/10.1670/10-267]

[196] A.J. Underwood, and M.G. Chapman, "Scales of spatial patterns of distribution of intertidal invertebrates", *Oecologia,* vol. 107, no. 2, pp. 212-224, 1996.
[http://dx.doi.org/10.1007/BF00327905] [PMID: 28307307]

[197] A.L. Moody, W.A. Thompson, B. De Bruijin, A.I. Houston, and J.D. Goss-Custard, "The Analysis of the Spacing of Animals, with an Example Based on Oystercatchers During the Tidal Cycle", *J. Anim. Ecol.,* vol. 66, no. 5, pp. 615-628, 1997.
[http://dx.doi.org/10.2307/5915]

[198] P. Rogerson, *Statistical methods for Geography.* Sage Publications: London, UK, 2001.
[http://dx.doi.org/10.4135/9781849209953]

[199] M-J. Fortin, and M. Dale, *Spatial Analysis: A guide for Ecologists.* Cambridge University Press: Cambridge, UK, 2005.

[200] B.D. Ripley, "The Second-Order Analysis of Stationary Point Processes", *J. Appl. Probab.,* vol. 13, no. 2, pp. 255-266, 1976.
[http://dx.doi.org/10.1017/S0021900200094328]

[201] P.J. Clark, and F.C. Evans, "Distance to nearest neighbor as a measure of spatial relationships in populations", *Ecology,* vol. 35, no. 4, pp. 445-453, 1954.
[http://dx.doi.org/10.2307/1931034]

[202] P.A. Moran, "Notes on continuous stochastic phenomena", *Biometrika,* vol. 37, no. 1-2, pp. 17-23, 1950.
[http://dx.doi.org/10.1093/biomet/37.1-2.17] [PMID: 15420245]

[203] L. Anselin, "Local Indicators of Spatial Association-LISA", *Geogr. Anal.,* vol. 27, no. 2, pp. 93-115, 1995.
[http://dx.doi.org/10.1111/j.1538-4632.1995.tb00338.x]

[204] S. Getzin, C. Dean, F. He, J.A. Trofymow, K. Wiegand, and T. Wiegand, "Spatial patterns and competition of tree species in a Douglas-fir chronosequence on Vancouver Island", *Ecography,* vol. 29, no. 5, pp. 671-682, 2006.
[http://dx.doi.org/10.1111/j.2006.0906-7590.04675.x]

[205] M.L. Wells, and A. Getis, "The spatial characteristics of stand structure in Pinus torreyana", *Plant Ecol.,* vol. 143, pp. 153-170, 1999.
[http://dx.doi.org/10.1023/A:1009866702320]

[206] L. Gray, and F. He, "Spatial point-pattern analysis for detecting density-dependent competition in a boreal chronosequence of Alberta", *For. Ecol. Manage.,* vol. 259, no. 1, pp. 98-106, 2009.
[http://dx.doi.org/10.1016/j.foreco.2009.09.048]

[207] D.L. Phillips, and J.A. MacMahon, "Competition and Spacing Patterns in Desert Shrubs", *J. Ecol.,* vol. 69, no. 1, pp. 97-115, 1981.
[http://dx.doi.org/10.2307/2259818]

[208] P. Haase, F.I. Pugnaire, S.C. Clark, and L.D. Incoll, "Spatial patterns in a two-tiered semi-arid shrubland in southeastern Spain", *J. Veg. Sci.,* vol. 7, pp. 527-534, 1996.
[http://dx.doi.org/10.2307/3236301]

[209] H.J. Schenk, C. Holzapfel, J.G. Hamilton, and B.E. Mahall, "Spatial ecology of a small desert shrub on adjacent geological substrates", *J. Ecol.,* vol. 91, pp. 383-395, 2003.
[http://dx.doi.org/10.1046/j.1365-2745.2003.00782.x]

[210] R.T. Forman, and L.E. Alexander, "Roads and their major ecological effects", *Annu. Rev. Ecol. Syst.,* vol. 1, no. 29, pp. 207-231, 1998.
[http://dx.doi.org/10.1146/annurev.ecolsys.29.1.207]

[211] B. Ruediger, "Rare carnivores and highways: moving into the 21st century", *ICOWET,* vol. 12, no. 9, pp. 10-16, 1998.

[212] I.F. Spellerberg, "Ecological effects of roads and traffic: a literature review", *Glob. Ecol. Biogeogr.,* no. 7, pp. 317-333, 1998.

[213] S.C. Trombulak, and C.A. Frissell, "Review of Ecological Effects of Roads on Terrestrial and Aquatic Communities", *Conserv. Biol.,* vol. 1, no. 14, pp. 18-30, 2000.
[http://dx.doi.org/10.1046/j.1523-1739.2000.99084.x]

[214] R.T. Forman, and R.D. Deblinger, "The Ecological Road-Effect Zone of a Massachusetts (U.S.A.) Suburban Highway", *Conserv. Biol.,* vol. 1, no. 14, pp. 36-46, 2000.
[http://dx.doi.org/10.1046/j.1523-1739.2000.99088.x]

[215] G.P. Clarke, P.C. White, and S. Harris, "Effects of roads on badger Meles meles populations in south-west England", *Biol. Conserv.,* vol. 2, no. 86, pp. 117-124, 1998.
[http://dx.doi.org/10.1016/S0006-3207(98)00018-4]

[216] N.C. Kline, and D.E. Swann, "Quantifying wildlife road mortality Saguaro National Park", *ICOWET,* vol. 12, no. 9, pp. 23-318, 1998.

[217] S. Hauer, H. Ansorge, and O. Zinke, "Mortality patterns of otters (Lutra lutra) from eastern Germany",

J. Zool. (Lond.), vol. 3, no. 256, pp. 361-368, 2002.
[http://dx.doi.org/10.1017/S0952836902000390]

[218] J.E. Malo, F. Suáez, and A. Díez, "Can we mitigate animal-vehicle accidents using predictive models?", *J. Appl. Ecol.,* vol. 4, no. 41, pp. 701-710, 2004.
[http://dx.doi.org/10.1111/j.0021-8901.2004.00929.x]

[219] D. Ramp, J. Caldwell, K.A. Edwards, D. Warton, and D.B. Croft, "Modelling of wildlife fatality hotspots along the Snowy Mountain Highway in New South Wales, Australia", *Biol. Conserv.,* vol. 4, no. 126, pp. 474-490, 2005.
[http://dx.doi.org/10.1016/j.biocon.2005.07.001]

[220] D. Ramp, V.K. Wilson, and D.B. Croft, "Assessing the impacts of roads in peri-urban reserves: Road-based fatalities and road usage by wildlife in the Royal National Park, New South Wales, Australia", *Biol. Conserv.,* vol. 3, no. 129, pp. 348-359, 2006.
[http://dx.doi.org/10.1016/j.biocon.2005.11.002]

[221] A. Seiler, "Predicting locations of moose-vehicle collisions in Sweden", *J. Appl. Ecol.,* vol. 2, no. 42, pp. 371-382, 2005.
[http://dx.doi.org/10.1111/j.1365-2664.2005.01013.x]

[222] J. VanDerWal, L.P. Shoo, C. Graham, and S.E. Williams, "Selecting pseudo-absence data for presence-only distribution modeling: How far should you stray from what you know?", *Ecol. Modell.,* vol. 220, pp. 589-594, 2009.
[http://dx.doi.org/10.1016/j.ecolmodel.2008.11.010]

CHAPTER 7

GPS Data Mining for Monitoring Community Mobility of Individuals

Sungsoon Hwang[1,*], Timothy Hanke[2] and Christian Evans[2]

[1] *Department of Geography, DePaul University, Chicago IL, USA*

[2] *Physical Therapy, Midwestern University, Downers Grove IL, USA*

Abstract: In the recent years, GPS and GIS have been increasingly used in health research as they can be used to measure individuals' movement and environmental exposure. Community mobility is an important aspect of the function and the environmental interaction of an individual. Indicators of community mobility, including important places visited and the number of trips made, can be extracted from raw GPS trajectory data using a trip detection algorithm with GIS. Those indicators of community mobility were used to monitor how stroke patients return to community and how they respond to rehabilitation treatments. GPS-based spatial footprints integrated with geospatial data such as air pollution, food access, and land use in a Geographical Information System (GIS) environment can help understand the environmental contexts of health behavior.

Keywords: Activity space, Big data, Community mobility, Daily mobility, Environmental exposure, Exposure assessment, GPS trajectory data, Geospatial health, Geographic Information Systems (GIS), Geospatial medicine, Health GIS, mHealth, Spatial data mining, Staypoint detection, Trip detection.

INTRODUCTION

Recently, there has been an increasing interest in spatially-based methods for elucidating the role of environmental exposure in shaping health and well-being [1 - 3]. In particular, historical spatial trajectories of individuals tracked through Global Positioning System (GPS) can improve exposure assessment. Advances in location acquisition technologies, sensor networks, and recent development in health care policies (*e.g.*, Affordable Care Act) have brought about (and will bring about) an explosion of a new type of health-related data including individual trajectory, environmental exposure, and electronic health records [4].

* **Corresponding author Sungsoon Hwang:** Department of Geography, DePaul University, Chicago IL, USA; Tel: +1 (773) 325-8668; Fax: +1 (773) 325-4590; E-mail: shwang9@depaul.edu

Ana Cláudia Teodoro (Ed.)

This change presents challenges for spatial analysis of health-related data in Geographic Information Systems (GIS) that have been traditionally focused on aggregate data (like census). The present time is ripe for more research on analytics of big data such as GPS trajectory data to demonstrate the potential of geospatial technologies (including GPS, GIS, and remote sensing) in improving health care.

An ability to move around community and participate in life situations-community mobility—is an important aspect of health and functioning according to the World Health Organization (WHO)'s International Classification of Functioning, Disability and Health (ICF) [5]. Clinicians can infer functional status of patients based on community mobility, in addition to other health records. Further, measures of community mobility can help healthcare providers understand how physical and social environment influence a persons' levels of physical activity, use of community resources, and feelings about community integration that have implications for health and well-being [6]. While levels of physical activity (measured as amount of time spent on vigorous to moderate physical activity) are widely used as a outcome measure, community mobility is of more relevance to some population who are not engaged in intense physical activity (like patients or the elderly). Similarly, some healthy subjects might exhibit low community mobility even if they are physically active under various circumstances (*e.g.*, unaware of community resources, lack personal transport, or lack social network). In other words, physical activity measures may not be adequate to fully capture one's life situations. Measures of community mobility can help to understand contexts of an activity that occurs.

Community mobility has been conventionally measured using self-report and clinical tests. Measures based on self-report are known to be largely inaccurate due to bias in recall and unreliable memory. Measures based on clinical tests are free of context as the test is conducted in a controlled environment. Given advent of GPS tracking and increased use of location-aware devices (such as a smartphone), it is now possible to accurately measure a sequence of location of individuals in a free living condition at a high temporal frequency. GPS trajectory data, a set of temporally sequenced records of an individual's location and time (x, y, t) obtained from a GPS receiver for a continuous period of time (like one day to life time), can be used to objectively quantify some aspects of community mobility.

In recent years, more and more health researchers have used GPS technology for measuring community mobility, physical activity, and environmental exposure often along with the use of accelerometer [7 - 11]. Thus far, studies that use GPS for health studies have been focused on demonstrating feasibility of GPS technology. The utility of GPS data depends on how to turn a voluminous and imperfect GPS data into something usable for analysis. Meanwhile, procedures for processing GPS data, such as trip detection, have been developed to measure travel behavior [12, 13]. Additionally, a great deal of research related to GPS data mining (or trajectory data mining) has been conducted, but research along those lines has rarely been conducted with health applications mind [14 - 17]. This chapter attempts to fill these gaps by applying findings from trajectory computing to measuring community mobility.

In this work, an automated procedure was developed to extract the number of stops made from individual GPS trajectories that are recorded continuously during a one-week monitoring period. A stop or staypoint is a part of a trajectory where the traveling object exhibits little movement for some time to conduct some activity (*e.g.*, shopping at a grocery store, and being at workplace) [18]. Detecting a stop serves as a means to segment raw GPS trajectory data into parsimonious and meaningful units, and thus help to deal with large volume of raw data. Although the number of stops doesn't measure all aspects of the community mobility, it can be used to infer how well subjects move around community and participate in life situations to some extent. The purpose of mining GPS data (*i.e.*, detecting the number of stops) in the context of this study is to monitor how patients after stroke return to community over time following rehabilitation treatments, and how well those patients respond to rehabilitation treatment. Detecting stops from raw GPS data is also useful in inferring significant places and activities conducted at those places if data is combined with geographical data like Point of Interest (POI) data in a GIS environment.

The remaining part of this chapter reviews recent health research that uses GPS, GPS-based outcome measures related to community mobility, and methods for GPS data processing relevant to community mobility measures. Then the algorithm for extracting community mobility measures from raw GPS data is described. The chapter discusses results on the evaluation of the algorithm and practical use of community mobility measures in relation to GIS.

USE OF GPS AND GIS IN HEALTH STUDIES

Determining where events like physical activity, diet, and exposure to health hazards occur can help specify the environmental influence on health outcomes. Until recently, it was difficult to measure the interplay between persons and the environment in relation to health outcomes. With GPS, it is possible to accurately quantify the location of physical activity. GPS data combined with geographical data (*e.g.*, land use, traffic, air pollution, and POIs) can help determine type (domain) of location where physical activity occurs. The use of GPS, geographical data, and accelerometer can improve the integrated measurement of health behavior in an environmental context, and further help elucidate the role of environmental exposure in health. GIS is used to integrate GPS-based location data, accelerometry-based activity data, and geographic data, and analyze environmental contexts of health behavior. Moreover, research along these lines have practical applications for personalizing physical activity interventions and informing policy changes for urban design that promotes healthy behavior. Advent of ubiquitous sensing on mobile platforms could improve exposure assessment and further applications of spatially based methods in health research. In the following were reviewed different ways in which GPS data and GIS analysis are used in health research.

Assessing Location of Physical Activity

In an effort to combat obesity, a lot of research has been conducted to determine where persons (mostly children) are physically active, using accelerometer and GPS devices. Several studies go further to identify type or domain of location where persons are engaged in physical activity (*e.g.*, leisure, school, transport, home) by overlaying location-specific physical activity data over remotely sensed images, land use data layers or POIs data in a GIS environment [19, 20]. Several studies found that being outdoors (sidewalk, school, playground, and active transport) are associated with high level of physical activity among children [21 - 25]. Children who walk to school were shown to be engaged in physical activity more than children who do not walk to school [26, 27]. Active commuting appears to be a viable means to increase physical activity. Overall findings suggest that interventions and policies for promoting physical activity should take into account environmental variables.

A growing number of researches have examined the role of the built environment on physical activity [28 - 30]. GIS techniques can be used to measure features of the built environment (such as residential density, land use mix, street connectivity, access to activity centers, and sidewalk safety) [31]. The idea is to determine whether the environment encourages or inhibits physical activity, and it is worthwhile considering policy changes in the physical environment in a geographically targeted manner. For instance, one can take into account research findings along these lines in choosing where to build sports facilities, where to add bike lanes and sidewalk, and determining how to enhance access to green space. In different studies, it is found that high population density and street connectivity are associated with high level of physical activity among adults [29, 32]. It appears that there is a growing consensus that targeted environmental physical activity interventions should be considered in conjunction with informational and behavioral approach to promoting physical activity [30, 33].

Monitoring Community Mobility

Clinicians often need to monitor community mobility to infer the functional status of those with physical and cognitive disability, in addition to performing gait test in the lab. This requires measurement of individuals' movement in a free living condition continuously using GPS devices and/or accelerometer. It was shown to be feasible to monitor community mobility of patients using GPS devices [10]. GPS might be considered to be a viable means to monitor how patients return to normal after rehabilitation treatments or surgery. Out-of-home mobility of some population (such as older adults) is often seen as an indicator of community integration and social interaction. Research shows that time spent outdoor and the amount of outdoor physical activity was positively associated with quality of life and cognitive functioning among older adults [34].

As GPS-based measures of community mobility are relatively new, there exist different approaches to quantifying community mobility. Community mobility was measured as the number of trips made, the number of targets (significant places provided by participants) reached, time spent out of the house, and trip mode [35, 36]. Hwang *et al.* [35] describe automated procedures for extracting the spatial extent of GPS track points out of home, the number of targets reached, and the number of trips from raw GPS trajectory data. Jayaraman *et al.* [36] present methods for measuring time spent out of the house and trip mode using GPS and accelerometer. It was noted that it is necessary to develop automated procedures that also treat uncertainty of a large volume of GPS trajectory data (*e.g.*, presence of outliers, gaps, and jitters), to create clinically practical tools. Types of locations visited or purpose of trips can be inferred more accurately if GPS trajectory data is overlaid with contextual data (such as transportation network and POI) in GIS.

Understanding Environmental Contexts of Health Behavior

An increasing number of studies employ integrated measurement of physical activity, the environmental context where an activity occurs, and diet intake, for holistic understanding of the health behavior-environment interaction. Zenk *et al.* [11] examine demographic and environmental (*e.g.*, land use, distance to fast food restaurants) features of neighborhoods and activity spaces of subjects using GPS, accelerometer, food diary, and GIS techniques. Fast food density in the daily path area (as a measure of activity space) was positively associated with saturated fat intake but supermarket availability in the activity space was not associated with dietary intake.

A mobile app, CalFit, records activity counts and energy expenditure and the time and location where an activity occurs using GPS devices and the accelerometer that are incorporated into smartphones [37]. This software can be also combined with dispersion models of ambient exposure to air pollutants [38]. Another mobile sensing platforms, the Personal Environmental Impact Report (PEIR) systems, allow participants to assess their exposures to fine particulate matter and fast food outlets and their impacts such as transportation-related greenhouse gas emissions [38]. PEIR relies on activity-classification systems (walking or driving) that uses data on location, speed, land use and traffic from GPS and GIS source data.

GPS-BASED MEASURES OF COMMUNITY MOBILITY

Potential for using GPS data in health research will grow as a large volume of individual GPS trajectory data is being generated from location-aware devices. GPS provides a means to continuously measure spatial footprint of individuals at high spatial and temporal resolution over an extended period of time. Several outcome measures that quantify community mobility using GPS data have been proposed. Activity space corresponds to one's exhaustive spatial footprints, and can be measured based on spatial dispersion of GPS track points. Aggregation of physical activity (eg, the number of steps) over various locations can also be used to measure physical activity and environmental interaction. Mobility of free-living individuals can be measured in terms of time spent out of home, average walking time/distance, the number of nodes (places) visited, and transportation corridors with environmental interaction. Below are reviewed different approaches to quantify community mobility from GPS data and analyze community mobility in GIS that can be used in health research.

Activity Space

Residential location has been traditionally identified as geographic space of significance in health research. This fact results in a mischaracterization of environmental exposures as there are other important locations. With the advent of location acquisition technology (like GPS, wireless network), it is possible to measure the geographic space in which activities have taken place (activity space). This allows for assessing environmental exposure beyond residential location (at multiple places) [1, 4]. Activity space consists of a set of spatial locations visited by an individual over a given period, and corresponds to her/his exhaustive spatial footprint.

Activity space is commonly measured as the standard deviation ellipse, the minimum convex polygon, and daily path area from GPS track points recorded for an individual during a given time duration (from one day to life time) [11, 39]. These measures of activity space basically represent the spatial dispersion of GPS track points. The axis of standard deviation ellipse (SDE) represents how large activity space is. The minimum convex polygon covers GPS track points exhaustively. Daily path area is the buffer from GPS track points that can be created using the appropriate GIS tools.

A limitation of this approach to measuring activity space is that GPS track points are assumed to carry the equal weight although track points carry varying weight due to irregular sampling rate and missing data. That is, some track points that comprise frequently visited locations should be treated differently from track points that make up a part of driving route (with less level of exposure). Similarly, level of environmental exposure to air pollution shouldn't be the same between all routes taken (tunnel, underground, public transit, and highway). GPS-based location data can be overlaid with other geospatial data (*e.g.*, land use, POIs, environmental data) to accurately assess types of locations or network paths visited using GIS techniques.

Movement and Activity in Physical Space Score (MAPS)

According to the WHO's International Classification of Functioning, Disability and Health (ICR), "function" is defined as the dynamic interaction of a persons' physical activity within his or her environment. This means that the function cannot be fully understood without comprehending the individual-environment interaction. Herrman *et al.* [40] proposed Movement and Activity in Physical Space Score (MAPS) to measure physical activity in an environmental context as a functional outcome measure. The score is calculated as the aggregation of physical activity per minute by locations other than home. MAPS is defined as:

$$MAPS = \sum_{L=1}^{n} \left(\frac{Activity}{Minutes} \right) \tag{1}$$

where L represents locations other than home. Out-of-home locations can be detected from raw GPS data. In eq. (1), *Activity* was measured as step count provided by the accelerometer, and *Minutes* is the number of minutes spent at the location for more than 10 minutes. Ten minutes was chosen to minimize misclassification of locations due to long stoplights and traffic congestion. A higher MAPS score can be interpreted as a higher level of function.

The formula presented in eq. (1) provides a way to examine variation in physical activity by different locations. No distinction, however, is made between high MAPS score with low environmental interaction (less locations visited) and high MAPS score with high environmental interaction (more locations visited). If locations visited are overlaid on land use data in GIS, types of activity (*e.g.*, occupational, recreational) can be inferred. In any case, accuracy of the MAPS score is sensitive to accuracy and reliability of locations (or staypoint) detection algorithm.

Out-of-home Mobility

Out-of-home mobility refers to the physical ability to move about outside the home, whether by foot or by any other means of transportation, and means the realization of all types of trips and activities outside the home [6]. Out-of-home activity was found to have positive impact on feeling of satisfaction and autonomy for people with mild and moderate dementia [41]. People who are more mobile report less loneliness and stronger feelings for community integration [6]. In a study of people with mild cognitive impairment, caregivers were found to be less burdened with care-recipients who are mobile outdoors [42].

Out-of-home mobility can be measured as variables calculated from GPS data, including average time spent out of home, average walking per day, average number of visited nodes, and average walking distance. Out-of-home mobility can help understand the interplay between personal competencies and resources as well as aspects of their physical and social environments. Using GIS, it is possible to identify features of the physical environments that encourages or inhibits physical activity (*e.g.*, access to green space, access to sports facilities, access to public transport, slope, presence of stairs, facilities unfriendly for people with disability).

Daily Mobility

Daily mobility refers to everyday movement of individuals over space between activity locations [43]. Daily mobility is of interest in environment-health research as a source of transport-related physical activity and as a vector of environmental exposure. By accounting for daily mobility patterns and activity space, public health practitioners can identify low-mobility population with access to low-resource/high-exposure environment.

Chaix *et al*. [43] suggest the use of "travel and activity place network area" to represent daily mobility. Travel and activity place network area consist of activity centers in which individuals spend a substantial portion of their time, optional destinations, and transportation corridors that allow exchange with the surrounding environment. Activity centers have important material and symbolic meanings to individuals, and are associated with a marker of willingness to do the behavior.

To make community mobility more meaningful for clinicians, significance of places visited, and the environmental context in which physical activity occurs (*e.g.*, park, gym, workplace, home, and bike path) need to be determined and incorporated into the mobility measure. GIS can be used to determine the context of physical activity. However, components of outcome measures, including the number of nodes visited and paths taken, can't be reliably obtained without proper GPS data processing techniques due to a large volume of GPS data with uncertainty.

GPS TRAJECTORY DATA MINING

Techniques for GPS trajectory data mining have been employed in different contexts. GPS trajectory was mined to detect trips and trip ends from individual trajectories to construct travel diaries or personal gazetteers that record time, location, and purposes of trips made and places visited [12, 13, 44]. Detecting significant or interesting places from location history compiled from a collection of trajectories is of interest in mining user interests and discovering pattern [45 - 48]. Once important places visited and routes taken are detected, activities can be inferred based on the intersection of those places/routes and contextual data such as POIs and transportation network [49, 50].

Overview of Trajectory Data Analysis

The field dedicated to trajectory data analysis (or trajectory computing) is a wide-ranging [14, 16, 51]. Techniques for trajectory computing can be classified into three categories: pre-processing, management, and mining [16]. Pre-processing is to make imperfect raw data clean, accurate, and concise. Management is concerned with efficiently accessing and retrieving data. Mining encompasses various techniques for uncovering patterns and behaviors. Trajectory computing can be applied to individual trajectory data [12, 13, 52] or collective trajectory data [48, 53]. Trajectory computing can be performed with or without contextual data (like POIs data). This section reviews related work that might be useful for automatically detecting trips or stops from individual GPS trajectory data without contextual data as they relate to measuring community mobility.

Due to a large volume of GPS data, it would be useful to partition an individual trajectory into meaningful segments. What constitutes "meaningful" segments depends on the context and purpose of analysis. In the current study that sets out to measure the number of trips made and places visited, we will focus on segmenting GPS trajectory into episodes of two types: stop episodes and move episodes [18, 51]. A stop (also known as staypoint, places, location, nodes) is a part of a trajectory where the traveling object exhibits little movement for greater than the defined minimum time duration. A move (also known as paths, trips, routes) is a part of trajectory that is not a stop, and is delimited by two consecutive stops.

The following reviews different approaches to detecting stops or moves from raw GPS data without contextual data, which is largely known as "trip detection algorithm". Although a focus is on automated procedures (algorithms) for trip detection, some research that employs methods that involve intervention is also reviewed if it can be adapted to the trip detection algorithm.

Rule-based Approach to Trip Detection

Stopher *et al*. [12] propose to detect trip ends where an individual makes a stop to conduct an activity such as work, and shop at a store. Stops are in essence where there is non-movement for some duration of time. Stops are where consecutive location change is less than 6 meters and the heading is unchanged and elapsed time is greater than or equal to 120 seconds. The authors also offer methods for correcting for GPS signal loss (or gaps). The speed for the gap is estimated as gap distance / (gap time - 2 minute). Then if the estimated speed matches both the average speed of the last 20 records of the previous trip and the average speed of the first 20 records of the following trip, a stop is considered to have occurred. Otherwise, no stop is considered to have occurred.

Schuessler & Axhausen [13] describe the procedure for detecting activities and trip modes from person-based raw GPS data without additional data. The procedure is divided into pre-processing, activity detection, and trip mode detection. Pre-processing includes deleting spatial outliers (that represent sudden jump in location with the maximal speed 50 m/s) and filtering speed values using the Gaussian kernel filtering with 15 second time range. Activities at stop locations are detected based on three criteria: low speed, high point density, and gap duration greater than 15 minutes. Trip mode is assigned to each segment using fuzzy rules based on speed and acceleration.

Li *et al*. [52] propose the algorithm that detects a staypoint automatically from raw data. A staypoint is subsequence of trajectories that appear to be stationary for some time. The algorithm identifies sub-trajectories that consist of two or more data points as a staypoint if the distance among two successive data points exceeds a distance threshold and the time span among those points exceeds a time threshold. The study uses 200 meters as a distance threshold, and 30 minutes as a time threshold.

Clustering-based Approach to Stop Detection

Ashbrook & Starner [45] present a method that automatically clusters GPS data taken over an extended period of time into meaningful locations at multiple scales. Spatial clusters are first detected using *k*-means clustering algorithm with a time threshold 10 minutes and a location radius of 0.5 mile. Then sublocation (meaningful locations) is detected within spatial clusters based on the optimal number of clusters. If elapsed time is greater than 10 minutes in any gap, then the gap is flagged as a meaningful location (stop). The authors offer some empirical grounds for choosing the time and distance threshold required for stop detection. 3 minutes, 10 minutes, 0.2 miles, and 0.5 miles are suggested.

A density-based spatial clustering algorithm can be used to detect clusters of arbitrary shape. In other words, spatially connected GPS track logs that span beyond a distance threshold can be detected using the clustering algorithm like DBSCAN [54]. DBSCAN detects a group of dense spatial clusters by aggregating spatial clusters that are density-reachable. DBSCAN begins with scanning the number of points within a specified bandwidth (*eps*) from any unvisited arbitrary data points, and checks whether the number of points exceeds a pre-specified threshold value (*MinPts*). If the above mentioned condition is met, the algorithm checks whether each point within the spatial cluster (identified above) forms another spatial cluster (*i.e.*, expandable). An advantage of density-based clustering algorithms is that they are relatively robust to noise, and can detect spatial clusters of arbitrary shape.

Yuan *et al.* [53] used the clustering algorithm to recommend where taxi drivers are likely to find customers by detecting parking places where taxi drivers are stationary to wait for customer pick-ups from collective GPS trajectories of vehicles. Hwang *et al.* [36] detect a staypoint by checking whether data points constituting a spatial cluster (detected by DBSCAN) meets temporal criteria (*i.e.*, exceeds the minimum required). Gaps are not taken into account in the staypoint detection in the study above. Additional features like change of direction and sinuosity can be also considered in the staypoint detection [48, 55].

If the scope is limited to trip detection based on raw data only, a staypoint or stop episode can usually be detected on the basis of one or more of the following features: gaps, speed, spatial density, and time duration. A position where there is no GPS signal (the GPS is off or the car enters tunnel) can be set to a staypoint [45]. A problem with this approach is that there is no guarantee that those gaps represent a stop episode. Similarly, GPS logs whose speed value is near zero can be identified as a stop episode [13]. This approach is limited because speed value is often inconsistent due to abnormalities present in GPS measurement. It seems that it is necessary to infer semantics of gaps (*i.e.*, do they represent stops or moves?), and reduce variability of inconsistent feature values through data cleaning techniques.

THE PROPOSED METHOD

This section presents the algorithm for mining measures of community mobility from raw GPS trajectories. The algorithm was applied to a new data set to demonstrate the practical use of GPS-based mobility measures. The input of the algorithm is individual GPS trajectory data that are recorded at a regular time interval r in seconds. The output of the algorithm is community mobility measures including the number of stops visited, the number of trips made, and the number of targets (important places provided by participants) reached. Stops are defined as locations where a traveling object stays for at least t (the minimum time duration for a stop) in seconds out of home. Other input parameters related to spatial density include *eps*, and *MinPts* for DBSCAN.

The framework consists of four modules: data cleaning, gap treatment, staypoint detection, and mobility measures extraction (Fig. **1**). The data cleaning module deletes spatial outliers. The gap treatment module fills gaps using a linear interpolation. The staypoint detection module identifies staypoints based on input parameters of spatial density and the minimal time duration using the adapted DBSCAN. A record that is flagged as a staypoint is classified as a stop, and a record that is not a staypoint is classified as a move. Any misclassified records are removed using the filtering technique. The mobility measures extraction module calculate community mobility measures from geographic objects representing a set of stop and move in a GIS environment. The current work extends the previous work [35], but treats gaps such that gaps can be classified into either stop or move. This is to address a limitation of density-based spatial clustering where a staypoint is not detected due to the insufficient number of data points in the presence of gaps. The entire process is automated in Python 2.7 and ArcPy site package of ESRI ArcGIS 10.3.

Data Cleaning

To check quality of raw GPS data, GPS track logs were overlaid on data of higher accuracy and independent source (ArcGIS 10.3 Map Service World Imagery). Most of GPS data were well aligned with reference data, but a few spatial outliers were present in trajectories recorded for a one-week period. Although location fix methods and Dilution of Precision could have been used to clean data, the method described in this chapter didn't consider such information for consistency because one of two GPS loggers we used for collecting data doesn't provide that information.

Data points were flagged as outliers if elapsed speed is much above normal values. A track log is deleted if the elapsed speed for consecutive data points is greater than 130 kilometer per hour. The elapsed speed is calculated as change in location between two consecutive data points divided by change in time. It is possible that false outliers (*e.g.*, overspeeding) can be deleted with this threshold, but the staypoint detection is not affected by false outliers and paths taken are constructed reliably despite those deleted false outliers.

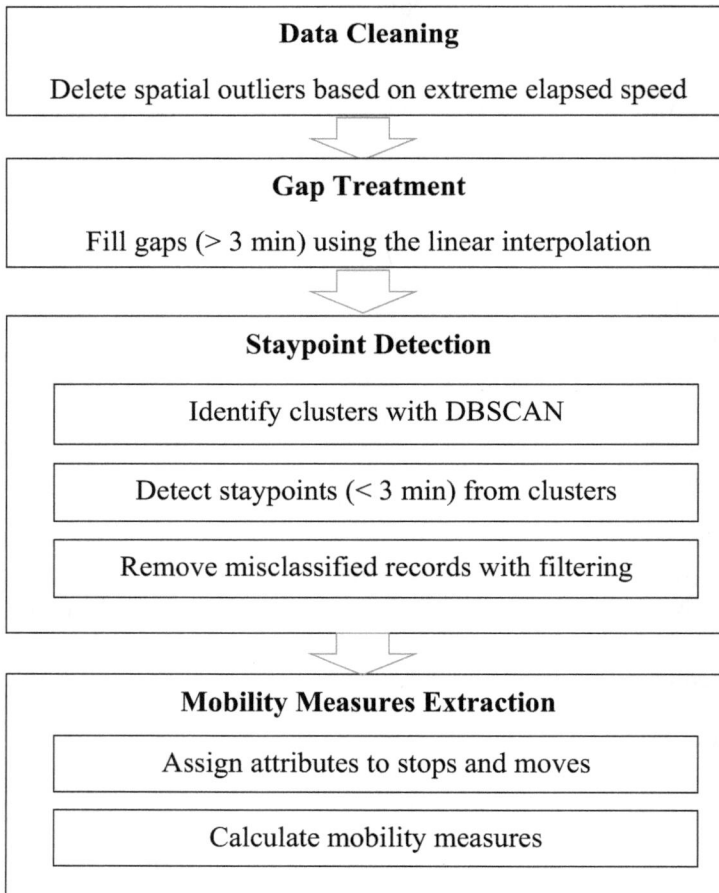

```
┌─────────────────────────────────────────────────────────┐
│                     Data Cleaning                         │
│                                                           │
│     Delete spatial outliers based on extreme elapsed speed│
└─────────────────────────────────────────────────────────┘
                            ⇩
┌─────────────────────────────────────────────────────────┐
│                     Gap Treatment                         │
│                                                           │
│        Fill gaps (> 3 min) using the linear interpolation │
└─────────────────────────────────────────────────────────┘
                            ⇩
┌─────────────────────────────────────────────────────────┐
│                   Staypoint Detection                     │
│   ┌───────────────────────────────────────────────────┐ │
│   │          Identify clusters with DBSCAN             │ │
│   └───────────────────────────────────────────────────┘ │
│   ┌───────────────────────────────────────────────────┐ │
│   │      Detect staypoints (< 3 min) from clusters     │ │
│   └───────────────────────────────────────────────────┘ │
│   ┌───────────────────────────────────────────────────┐ │
│   │     Remove misclassified records with filtering    │ │
│   └───────────────────────────────────────────────────┘ │
└─────────────────────────────────────────────────────────┘
                            ⇩
┌─────────────────────────────────────────────────────────┐
│                Mobility Measures Extraction               │
│   ┌───────────────────────────────────────────────────┐ │
│   │         Assign attributes to stops and moves       │ │
│   └───────────────────────────────────────────────────┘ │
│   ┌───────────────────────────────────────────────────┐ │
│   │           Calculate mobility measures              │ │
│   └───────────────────────────────────────────────────┘ │
└─────────────────────────────────────────────────────────┘
```

Fig. (1). Overview of the proposed method.

Gap Treatment

The gap treatment module adds a given number of data points (k) for a gap whose time duration exceeds a certain threshold (q), where parameters k and q are calculated as a function of input parameters, including *MinPts*, t, and r. More specifically, k data points are added just before any gap whose elapsed time from the previous track log is at least q seconds, where $k = MinPts + 1$, and $q = t + r$ where *MinPts* is the minimum number of points required for defining spatial density in DBSCAN, t is the minimum time duration in seconds of a staypoint and r is a recording time interval in seconds used for GPS logging. In this study, k is 6 and q is 210 seconds as t is 180 seconds and r is 30 seconds. Time and position of those data points added are linearly interpolated between data points before and after a qualified gap (> 210 seconds in this study).

The idea is that if the spatial distance between a data point before the gap (p_1) and a data point after the gap (p_2) is sufficiently large (*i.e.*, the distance is greater than *eps* and p_i and p_j are not density connected), then k data points that are added are unlikely to be detected as a staypoint due to low density. A good example is that an individual takes subway, a vehicle enters a tunnel, or GPS logger runs out of battery while moving. If the spatial distance between p_1 and p_2 is sufficiently small (*i.e.*, the distance is less than *eps,* and those p_1 and p_2 are density connected), then k data points that are added are likely to be detected as a staypoint. A good example is that GPS satellite signals are lost when one enters a building and the subsequent data point after the gap is relatively close to the previous data point, which most likely indicates that a stop is made at a place.

Staypoint Detection

For DBSCAN, *eps* (search radius) is set to 50 meters, and *MinPts* (minimum number of points) is set to 5 based on observed spatial accuracy of data and extent of spatial clusters. Obviously, *eps* and *MinPts* should be chosen in relation to t (the minimum time duration of a staypoint) and r (recording time interval). Spatial clusters are identified from DBSCAN given parameters *MinPts* and *eps*. With DBSCAN, track logs that are scattered around a staypoint are treated as noise, and thus those track logs (such as beginning of a new trip before or after making a stop) do not form part of a staypoint. Home locations are geocoded based on street addresses provided by participants in a GIS environment. Any data points within 50 meters from home locations are excluded for a staypoint detection to reduce processing time.

Three minutes is chosen as t because this allows for capturing activities for the short duration (such as running an errand). Previous studies and observation of the study area indicate that the minimum time duration for staypoints or important places is 2-30 minutes [12, 46, 47]. Then it was checked to see if track logs constituting a spatial cluster are consecutive for the minimal duration of time t. If the condition above is met, a spatial cluster is disaggregated into one or more stay points. So a place visited more than once are identified as multiple staypoints with different time stamps. Track logs that are identified as a staypoint are flagged as "stop", and track logs that are not identified as a staypoint are flagged as "move".

Some track logs can still remain misclassified after the staypoint detection module. For instance, anomalies in GPS measurement (such as indoor jitter) cause some track logs that are actually part of a staypoint to be classified as "move" due to low density around those spatial outliers. Conversely, track logs that are not semantically a stop episode (such as waiting for a traffic light for longer than t) can be falsely classified as "stop". What is common in those misclassified track logs is that they are surrounded by track logs that are classified otherwise. Filtering stop/move values in the moving window (or temporal neighbors) can fix this problem. In this study, stop/move values are replaced with the most common value of five consecutive track logs. The size of temporal neighbors is calculated as $t/r - 1$ (that is, 5 in this study) to make filtering results balanced between too noisy and too smooth outcomes.

Mobility Measures Extraction

Unique identifiers (IDs) are assigned to stops and moves based on temporal order and a rule that a stop is followed by a move, and vice versa. In operational terms, a stop is a sequence of track logs that are spatially clustered and temporally continuous; a move is a sequence of track logs that are not spatially clustered and temporally continuous. A move is delimited by two stops. Thus the number of stops will be matched with that of moves. Using functions of ArcPy site packages in a GIS environment, the module creates the mean center of track logs that are marked as a stop with the unique ID to reduce those track logs into a point geographic object. Similarly, the module converts track logs that are marked as a move into a line (or route) geographic object for data reduction. Then the module extracts characteristics of a point and line object that represents a stop and move, respectively. For instance, the point object representing a stop has coordinates and street address of representative location (using reverse geocoding), time duration, beginning time, and ending time. The geographic object representing a move possesses average speed, distance, and identifiers of origin and destination stops.

Potential measures of community mobility are calculated from two geographic objects (stops and moves) created above. They include the number of stops visited out of the house, number of moves (trips) made out of the house, duration of those stops, duration of those trips, length of those trips, and average speed of those trips. The number of out-of-home stops is matched with the number of out-o--home trips because each trip is forced to end at a stop. It was assumed that one move segment delimited by origin and destination stops is associated with one mode of transportation. If average speed of trips is greater than 8 kilometer per hour, then the mode of trips is set to driving. Otherwise, the mode of trips is set to walking.

In addition to automatically detecting stops and trips from raw GPS data, we also asked study participants to identify ten important places out of home they would like to go, called targets, before the study. Targets are geocoded (address-matched), and whether targets are reached was determined based on proximity in a GIS environment. That is, if GPS track points are within 150 meters from targets, then targets are marked to be reached. The module also calculates the percent of track points within multiple buffers from home at 50 meter (stay home), 1000 meter (neighborhood level), 5000 meter (town level), 20000 (out of town level), and beyond, respectively in GIS. Measures described above are intended to represent community participation and physical activity levels of subjects, and can be used to gauge levels of community mobility of subjects.

RESULTS

The automated procedure was applied to data collected for the study described in Evans *et al.* [10] to monitor how patients after stroke recover and return to community after rehabilitation treatment. Subjects are provided an informed consent to participate in the study, involving carrying a GPS logger (GlobalSat DG-100 Data Logger or QStarz Travel recorder XT) during waking hours of one week for this study. One week was chosen as a monitoring period because similar types of activities are repeated over one-week period, which allows for capturing a minimally distinct set of activities. Recording interval (r) was set to 30 seconds, and data were collected in a passive and continuous mode. Therefore, one GPS trajectory data is comprised of track logs continuously recorded for one week.

Data collection was collected from May 2009 to April 2015. We collect data from two sets of subjects—stroke patients who received rehabilitation treatment, and control subjects—from six milestones—1 week, 5 weeks, 9 weeks, 6 months (26 weeks), and 1 year (52 weeks)—after a baseline time (such as time of rehabilitation treatment). We have thus far collected total 81-week worth of data from 17 subjects (6 control subjects, and 14 rehab subjects). Among them 4 control subjects and 7 rehab subjects completed monitoring for the entire periods (six milestones above).

To evaluate the accuracy of the proposed staypoint detection algorithm, research assistants visually inspected track logs of nine trajectories (11% sample) from one control subject and one rehab subject against remotely sensed images on Google Earth in November 2014. One trajectory is one-week worth of data. This led to the compilation of the number of stops (# stops) validated for each of nine data set. Then the number of stops detected from the algorithm was compared to the number of stops validated. Fig. (**2**) shows that # stops detected by the algorithm is fairly well matched with # stops validated. The Pearson's correlation coefficient between # stops validated and # stops detected is .692 with *p*-value .039.

Table **1** shows # stops made by rehab subjects over six time periods (week 1, week 5, week 9, week 26, and week 52 following rehabilitation treatment) based on the proposed algorithm. Over the period of a year following rehabilitation treatments, community mobility exhibited by seven rehab subjects has increased. The average # stops made per day increased from 2.24 to 4.81.

Comparison of # stops detected against # stops validated

stops detected by the current work

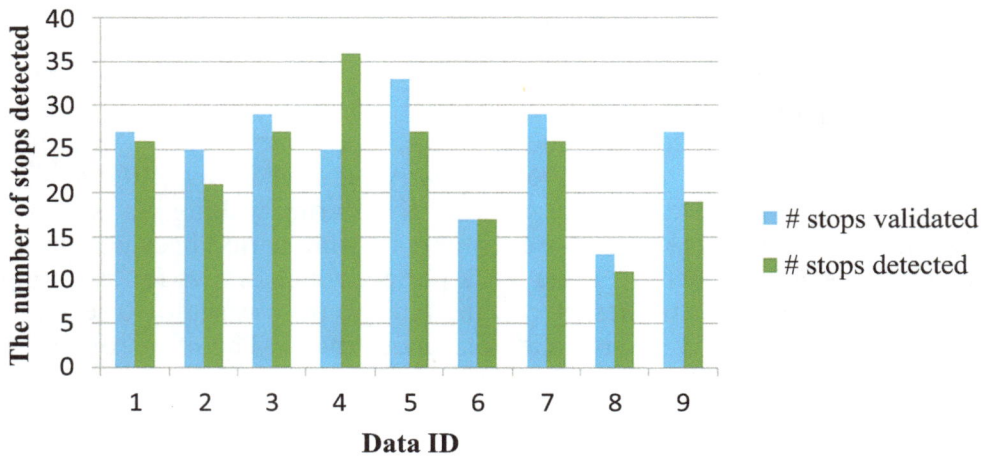

Fig. (2). Comparison of # stops detected against # stops validated.

Individual difference can be noted in Table **1**. The number of stops made by Subject 01 and 12 has fluctuated during the monitoring period while other rehab subjects exhibit a steady increase in the number of stops made. This individual difference can inform health care professionals regarding efficacy of rehabilitation treatments at an individual level, and help them personalize treatments.

Table 1. The number of stops made by rehab subjects over six time periods.

Subject	W01	W05	W09	W26	W52
01	33	21	25	12	23
03	13	8	18	16	16
04	21	26	26	55	63
05	7	6	5	10	7

(Table 1) contd.....

Subject	W01	W05	W09	W26	W52
08	12	7	32	40	58
09	6	27	20	24	35
12	18	42	24	21	34
Average (per week)	15.71	19.57	21.43	25.43	33.71
Average (per day)	2.24	2.79	3.06	3.63	4.81

Control subjects made on average 6.75 to 5.61 stops per day, which is significantly more than rehab subjects made (Table **2**). Furthermore, # stops made by control subjects doesn't exhibit much pattern (a minor decrease) during the monitoring period in contrast to # stops made by rehab subjects. This confirms that stroke patients' community mobility has increased following rehabilitation treatments.

Table 2. The number of stops made by control subjects over six time periods.

Subject	W01	W05	W09	W26	W52
01	40	15	21	19	32
03	30	26	35	42	27
05	52	58	58	47	42
06	67	77	59	53	56
Average (per week)	47.25	44	43.25	40.25	39.25
Average (per day)	6.75	6.29	6.18	5.75	5.61

Lower levels of community mobility by rehab subjects in comparison to control subjects are also substantiated by the tabulation of the number of targets reached. Rehab subjects reached about 3 to 4 targets on average whereas control subjects reached 7 to 8 targets on average during the monitoring period as seen in Table **3** and **4**.

Table 3. The number of targets reached by rehab subjects over six time periods.

Subject	W01	W05	W09	W26	W52
01	5	2	5	2	4
03	5	3	2	5	5
04	6	7	4	6	5
05	4	2	3	4	3
08	4	4	2	4	4
09	4	5	5	4	4

(Table 3) contd.....

Subject	W01	W05	W09	W26	W52
12	6	5	3	5	3
average	4.86	4	3.43	4.29	4

Table 4. The number of targets reached by control subjects over six time periods.

Subject	W01	W05	W09	W26	W52
01	8	7	8	8	8
03	4	3	5	2	4
05	10	10	10	10	9
06	10	9	8	9	10
average	8	7.25	7.75	7.25	7.75

CONCLUDING REMARKS

This chapter shows how the proposed trip detection algorithm can be used to derive indicators of community mobility from raw GPS trajectory data. The trip detection algorithm that takes into account gaps is proposed in this chapter. GIS techniques including proximity analysis, overlay, geocoding, spatial statistics, and data manipulation, are used to process GPS data in the algorithm. The algorithm automatically extracts indicators of community mobility, such as the number of out-of-home trips, the number of important places (target) visited, and spatial dispersion of subjects' footprint, from raw GPS trajectory data.

If the processed GPS trajectory data are integrated with geospatial data (*e.g.*, land use, environmental & demographic data, POI), one can discover individuals' life pattern and better understand environmental influences on health and well-being. Community mobility measures of stroke patients demonstrated in this case study, can help health care professionals infer patients' functional status and monitor how patients respond to clinical intervention. This chapter demonstrates that GPS/GIS integrated methods presented in this chapter can be used to improve the exposure assessment for elucidating role of environmental exposure in health outcomes.

CONSENT FOR PUBLICATION

Not applicable.

CONFLICT OF INTEREST

The author (editor) declares no conflict of interest, financial or otherwise.

ACKNOWLEDGEMENTS

Declared none

REFERENCES

[1]   D.B. Richardson, N.D. Volkow, M-P. Kwan, R.M. Kaplan, M.F. Goodchild, and R.T. Croyle, "Medicine. Spatial turn in health research", *Science,* vol. 339, no. 6126, pp. 1390-1392, 2013.
[http://dx.doi.org/10.1126/science.1232257] [PMID: 23520099]

[2]   National Research Council, *Exposure Science in the 21st Century: a Vision and a Strategy.* National Academies Press, 2012.

[3]   C.P. Wild, "Complementing the genome with an "exposome": the outstanding challenge of environmental exposure measurement in molecular epidemiology", *Cancer Epidemiol. Biomarkers Prev.,* vol. 14, no. 8, pp. 1847-1850, 2005.
[http://dx.doi.org/10.1158/1055-9965.EPI-05-0456] [PMID: 16103423]

[4]   A.J. Blatt, *Health, Science, and Place: A New Model.* Springer, 2014, pp. 11-21.

[5]   World Health Organization, *International Classification of Functioning, Disability and Health: ICF.* World Health Organization, 2001.

[6]   H. Mollenkopf, *Enhancing Mobility in Later Life: Personal Coping, Environmental Resources and Technical Support; the Out-of-home Mobility of Older Adults in Urban and Rural Regions of Five European Countries: IOS Press*, 2005.

[7]   J. Kerr, S. Duncan, and J. Schipperijn, "Using global positioning systems in health research: a practical approach to data collection and processing", *Am. J. Prev. Med.,* vol. 41, no. 5, pp. 532-540, 2011.
[http://dx.doi.org/10.1016/j.amepre.2011.07.017] [PMID: 22011426]

[8]   M.M. Jankowska, J. Schipperijn, and J. Kerr, "A framework for using GPS data in physical activity and sedentary behavior studies", *Exerc. Sport Sci. Rev.,* vol. 43, no. 1, pp. 48-56, 2015.
[http://dx.doi.org/10.1249/JES.0000000000000035] [PMID: 25390297]

[9]   P.J. Krenn, S. Titze, P. Oja, A. Jones, and D. Ogilvie, "Use of global positioning systems to study physical activity and the environment: a systematic review", *Am. J. Prev. Med.,* vol. 41, no. 5, pp. 508-515, 2011.
[http://dx.doi.org/10.1016/j.amepre.2011.06.046] [PMID: 22011423]

[10]   C.C. Evans, T.A. Hanke, D. Zielke, S. Keller, and K. Ruroede, "Monitoring community mobility with global positioning system technology after a stroke: a case study", *J. Neurol. Phys. Ther.,* vol. 36, no. 2, pp. 68-78, 2012.
[http://dx.doi.org/10.1097/NPT.0b013e318256511a] [PMID: 22592062]

[11]   S.N. Zenk, A.J. Schulz, S.A. Matthews, A. Odoms-Young, J. Wilbur, L. Wegrzyn, K. Gibbs, C. Braunschweig, and C. Stokes, "Activity space environment and dietary and physical activity behaviors: a pilot study", *Health Place,* vol. 17, no. 5, pp. 1150-1161, 2011.
[http://dx.doi.org/10.1016/j.healthplace.2011.05.001] [PMID: 21696995]

[12]   P. Stopher, C. FitzGerald, and J. Zhang, "Search for a global positioning system device to measure person travel", *Transp. Res., Part C Emerg. Technol.,* vol. 16, pp. 350-369, 2008.
[http://dx.doi.org/10.1016/j.trc.2007.10.002]

[13]   N. Schuessler, and K. Axhausen, "Processing Raw Data from Global Positioning Systems Without Additional Information", *Transp. Res. Rec.,* no. 2105, pp. 28-36, 2009.
[http://dx.doi.org/10.3141/2105-04]

[14]   Y. Zheng, X. Zhou, Ed., *Computing with spatial trajectories.* Springer Science & Business Media, 2011.
[http://dx.doi.org/10.1007/978-1-4614-1629-6]

[15]   M. Buchin, A. Driemel, M. van Kreveld, and V. Sacristan, "Segmenting trajectories: A framework and algorithms using spatiotemporal criteria", *Journal of Spatial Information Science.,* vol. 0, pp. 33-63, 2015.

[16]   Y. Zheng, "Trajectory Data Mining: An Overview", *ACM Trans. Intell. Syst. Technol.,* vol. 6, pp. 1-41, 2015.
[http://dx.doi.org/10.1145/2743025]

[17]   J.A. Carlson, M.M. Jankowska, K. Meseck, S. Godbole, L. Natarajan, F. Raab, B. Demchak, K. Patrick, and J. Kerr, "Validity of PALMS GPS scoring of active and passive travel compared with SenseCam", *Med. Sci. Sports Exerc.,* vol. 47, no. 3, pp. 662-667, 2015.
[http://dx.doi.org/10.1249/MSS.0000000000000446] [PMID: 25010407]

[18]   S. Spaccapietra, C. Parent, M.L. Damiani, J.A. de Macedo, F. Porto, and C. Vangenot, "A conceptual view on trajectories", *Data Knowl. Eng.,* vol. 65, pp. 126-146, 2008.
[http://dx.doi.org/10.1016/j.datak.2007.10.008]

[19]   C.D. Klinker, J. Schipperijn, H. Christian, J. Kerr, A.K. Ersbøll, and J. Troelsen, "Using accelerometers and global positioning system devices to assess gender and age differences in children's school, transport, leisure and home based physical activity", *Int. J. Behav. Nutr. Phys. Act.,* vol. 11, p. 8, 2014.
[http://dx.doi.org/10.1186/1479-5868-11-8] [PMID: 24457029]

[20]   N.M. Oreskovic, J.M. Perrin, A.I. Robinson, J.J. Locascio, J. Blossom, M.L. Chen, J.P. Winickoff, A.E. Field, C. Green, and E. Goodman, "Adolescents' use of the built environment for physical activity", *BMC Public Health,* vol. 15, p. 251, 2015.
[http://dx.doi.org/10.1186/s12889-015-1596-6] [PMID: 25880654]

[21]   A.P. Jones, E.G. Coombes, S.J. Griffin, and E.M. van Sluijs, "Environmental supportiveness for physical activity in English schoolchildren: a study using Global Positioning Systems", *Int. J. Behav. Nutr. Phys. Act.,* vol. 6, p. 42, 2009.
[http://dx.doi.org/10.1186/1479-5868-6-42] [PMID: 19615073]

[22]   R. Quigg, A. Gray, A.I. Reeder, A. Holt, and D.L. Waters, "Using accelerometers and GPS units to identify the proportion of daily physical activity located in parks with playgrounds in New Zealand children", *Prev. Med.,* vol. 50, no. 5-6, pp. 235-240, 2010.
[http://dx.doi.org/10.1016/j.ypmed.2010.02.002] [PMID: 20153361]

[23]   N.M. Oreskovic, J. Blossom, A.E. Field, S.R. Chiang, J.P. Winickoff, and R.E. Kleinman, "Combining global positioning system and accelerometer data to determine the locations of physical activity in children", *Geospat. Health,* vol. 6, no. 2, pp. 263-272, 2012.
[http://dx.doi.org/10.4081/gh.2012.144] [PMID: 22639128]

[24]   T.M. O'Connor, E. Cerin, J. Robles, R.E. Lee, J. Kerr, N. Butte, J.A. Mendoza, D. Thompson, and T. Baranowski, "Feasibility study to objectively assess activity and location of Hispanic preschoolers: a short communication", *Geospat. Health,* vol. 7, no. 2, pp. 375-380, 2013.
[http://dx.doi.org/10.4081/gh.2013.94] [PMID: 23733298]

[25]   L. Yin, S. Raja, X. Li, Y. Lai, L. Epstein, and J. Roemmich, "Neighbourhood for Playing: Using GPS, GIS and Accelerometry to Delineate Areas within which Youth are Physically Active", *Urban Stud.,* vol. 50, pp. 2922-2939, 2013.
[http://dx.doi.org/10.1177/0042098013482510]

[26]   A.R. Cooper, A.S. Page, B.W. Wheeler, P. Griew, L. Davis, M. Hillsdon, and R. Jago, "Mapping the walk to school using accelerometry combined with a global positioning system", *Am. J. Prev. Med.,* vol. 38, no. 2, pp. 178-183, 2010.
[http://dx.doi.org/10.1016/j.amepre.2009.10.036] [PMID: 20117574]

[27]   C. Lee, and L. Li, "Demographic, physical activity, and route characteristics related to school transportation: an exploratory study", *Am. J. Health Promot.,* vol. 28, no. 3, suppl. Suppl., pp. S77-S88, 2014.

[http://dx.doi.org/10.4278/ajhp.130430-QUAN-211] [PMID: 24380470]

[28]   E. Cerin, E. Leslie, N. Owen, and A. Bauman, Applying GIS in Physical Activity Research: Community 'Walkability' and Walking Behaviors.*GIS for Health and the Environment. Lecture Notes in Geoinformation and Cartography: Springer Berlin Heidelberg.,* P.C. Lai, A.S. Mak, Eds., , 2007, pp. 72-89.
[http://dx.doi.org/10.1007/978-3-540-71318-0_6]

[29]   D. Ding, and K. Gebel, "Built environment, physical activity, and obesity: what have we learned from reviewing the literature?", *Health Place,* vol. 18, no. 1, pp. 100-105, 2012.
[http://dx.doi.org/10.1016/j.healthplace.2011.08.021] [PMID: 21983062]

[30]   J.F. Sallis, M.F. Floyd, D.A. Rodríguez, and B.E. Saelens, "Role of built environments in physical activity, obesity, and cardiovascular disease", *Circulation,* vol. 125, no. 5, pp. 729-737, 2012.
[http://dx.doi.org/10.1161/CIRCULATIONAHA.110.969022] [PMID: 22311885]

[31]   L.E. Thornton, J.R. Pearce, and A.M. Kavanagh, "Using Geographic Information Systems (GIS) to assess the role of the built environment in influencing obesity: a glossary", *Int. J. Behav. Nutr. Phys. Act.,* vol. 8, p. 71, 2011.
[http://dx.doi.org/10.1186/1479-5868-8-71] [PMID: 21722367]

[32]   D.A. Rodríguez, A.L. Brown, and P.J. Troped, "Portable global positioning units to complement accelerometry-based physical activity monitors", *Med. Sci. Sports Exerc.,* vol. 37, no. 11, suppl. Suppl., pp. S572-S581, 2005.
[http://dx.doi.org/10.1249/01.mss.0000185297.72328.ce] [PMID: 16294120]

[33]   E.B. Kahn, L.T. Ramsey, R.C. Brownson, G.W. Heath, E.H. Howze, K.E. Powell, E.J. Stone, M.W. Rajab, and P. Corso, "The effectiveness of interventions to increase physical activity. A systematic review", *Am. J. Prev. Med.,* vol. 22, no. 4, suppl. Suppl., pp. 73-107, 2002.
[http://dx.doi.org/10.1016/S0749-3797(02)00434-8] [PMID: 11985936]

[34]   J. Kerr, S. Marshall, S. Godbole, S. Neukam, K. Crist, K. Wasilenko, S. Golshan, and D. Buchner, "The relationship between outdoor activity and health in older adults using GPS", *Int. J. Environ. Res. Public Health,* vol. 9, no. 12, pp. 4615-4625, 2012.
[http://dx.doi.org/10.3390/ijerph9124615] [PMID: 23330225]

[35]   S. Hwang, T. Hanke, and C. Evans, Automated Extraction of Community Mobility Measures from GPS Stream Data Using Temporal DBSCAN.*Computational Science and Its Applications – ICCSA 2013.,* B. Murgante, S. Misra, M. Carlini, C. Torre, H-Q. Nguyen, D. Taniar, Eds., vol. 7972. Springer Berlin Heidelberg, 2013, pp. 86-98.Lecture Notes in Computer Science
[http://dx.doi.org/10.1007/978-3-642-39643-4_7]

[36]   A. Jayaraman, S. Deeny, Y. Eisenberg, G. Mathur, and T. Kuiken, "Global position sensing and step activity as outcome measures of community mobility and social interaction for an individual with a transfemoral amputation due to dysvascular disease", *Phys. Ther.,* vol. 94, no. 3, pp. 401-410, 2014.
[http://dx.doi.org/10.2522/ptj.20120527] [PMID: 24092905]

[37]   A. de Nazelle, E. Seto, D. Donaire-Gonzalez, M. Mendez, J. Matamala, M.J. Nieuwenhuijsen, and M. Jerrett, "Improving estimates of air pollution exposure through ubiquitous sensing technologies", *Environ. Pollut.,* vol. 176, pp. 92-99, 2013.
[http://dx.doi.org/10.1016/j.envpol.2012.12.032] [PMID: 23416743]

[38]   M. Mun, "Reddy S, Shilton K, Yau N, Burke J, Estrin D, Hansen M, Howard E, West R, Boda P. PEIR, the Personal Environmental Impact Report, as a Platform for Participatory Sensing Systems Research. Mobisys 2009", *Proceedings of the 7th International Conference On Mobile Systems,* 2009pp. 55-68

[39]   J.A. Hirsch, M. Winters, P. Clarke, and H. McKay, "Generating GPS activity spaces that shed light upon the mobility habits of older adults: a descriptive analysis", *Int. J. Health Geogr.,* vol. 13, p. 51, 2014.
[http://dx.doi.org/10.1186/1476-072X-13-51] [PMID: 25495710]

[40]   S.D. Herrmann, E.M. Snook, M. Kang, C.B. Scott, M.G. Mack, T.P. Dompier, and B.G. Ragan, "Development and validation of a movement and activity in physical space score as a functional outcome measure", *Arch. Phys. Med. Rehabil.*, vol. 92, no. 10, pp. 1652-1658, 2011.
[http://dx.doi.org/10.1016/j.apmr.2011.05.001] [PMID: 21872844]

[41]   A. Phinney, H. Chaudhury, and D.L. O'Connor, "Doing as much as I can do: the meaning of activity for people with dementia", *Aging Ment. Health,* vol. 11, no. 4, pp. 384-393, 2007.
[http://dx.doi.org/10.1080/13607860601086470] [PMID: 17612802]

[42]   S. Werner, G.K. Auslander, N. Shoval, T. Gitlitz, R. Landau, and J. Heinik, "Caregiving burden and out-of-home mobility of cognitively impaired care-recipients based on GPS tracking", *Int. Psychogeriatr.,* vol. 24, no. 11, pp. 1836-1845, 2012.
[http://dx.doi.org/10.1017/S1041610212001135] [PMID: 22874772]

[43]   B. Chaix, Y. Kestens, C. Perchoux, N. Karusisi, J. Merlo, and K. Labadi, "An interactive mapping tool to assess individual mobility patterns in neighborhood studies", *Am. J. Prev. Med.,* vol. 43, no. 4, pp. 440-450, 2012.
[http://dx.doi.org/10.1016/j.amepre.2012.06.026] [PMID: 22992364]

[44]   C. Zhou, D. Frankowski, P. Ludford, S. Shekhar, L. Terveen, Ed., *Discovering Personal Gazetteers: An Interactive Clustering Approach2004.* ACM: New York, NY, USA, 2004.
[http://dx.doi.org/10.1145/1032222.1032261]

[45]   D. Ashbrook, and T. Starner, "Learning significant locations and predicting user movement with GPS", *Sixth International Symposium on Wearable Computers, 2002 (ISWC 2002) Proceedings; IEEE,* 2002pp. 101-108
[http://dx.doi.org/10.1109/ISWC.2002.1167224]

[46]   Y. Ye, Y. Zheng, Y. Chen, J. Feng, and X. Xie, "Mining Individual Life Pattern Based on Location History", *Tenth International Conference on Mobile Data Management: Systems, Services and Middleware,* 2009pp. 1-10
[http://dx.doi.org/10.1109/MDM.2009.11]

[47]   "A Clustering-based Approach for Discovering Interesting Places in Trajectories", *Proceedings of the 2008 ACM symposium on Applied Computing,* 2008pp. 863-868
[http://dx.doi.org/10.1145/1363686.1363886]

[48]   "DB-SMoT: A direction-based spatio-temporal clustering method", *Intelligent Systems (IS), 2010 5th IEEE International Conference. IEEE,* 2010pp. 114-119

[49]   L. Liao, D. Fox, and H. Kautz, "Extracting Places and Activities from GPS Traces Using Hierarchical Conditional Random Fields", *Int. J. Robot. Res.,* vol. 26, pp. 119-134, 2007.
[http://dx.doi.org/10.1177/0278364907073775]

[50]   A. Rodrigues, C. Damásio, and J.E. Cunha, Using GPS Logs to Identify Agronomical Activities.*Connecting a Digital Europe Through Location and Place. Lecture Notes in Geoinformation and Cartography.,* J. Huerta, S. Schade, C. Granell, Eds., Springer International Publishing, 2014, pp. 105-121.
[http://dx.doi.org/10.1007/978-3-319-03611-3_7]

[51]   C Parent, S Spaccapietra, C Renso, G Andrienko, N Andrienko, and V Bogorny, "Semantic Trajectories Modeling and Analysis", *ACM Comput Surv.,* vol. 45, no. 42, pp. 1-32, 2013.
[http://dx.doi.org/10.1145/2501654.2501656]

[52]   Q. Li, Y. Zheng, X. Xie, Y. Chen, W. Liu, and W-Y. Ma, "Mining User Similarity Based on Location History", *Proceedings of the 16th ACM SIGSPATIAL international conference on Advances in geographic information systems,* 2008pp. 34-43
[http://dx.doi.org/10.1145/1463434.1463477]

[53]   N.J. Yuan, Y. Zheng, L. Zhang, and X. Xie, "T-Finder: A Recommender System for Finding Passengers and Vacant Taxis", *IEEE Trans. Knowl. Data Eng.,* vol. 25, pp. 2390-2403, 2013.

[http://dx.doi.org/10.1109/TKDE.2012.153]

[54]   M. Ester, H-P.S.J. Kriegel, and X. Xu, "A density-based algorithm for discovering clusters in large spatial databases with noise", *Proceedings of the second international conference on knowledge discovery and data mining (KDD '96)*, 1996pp. 226-231

[55]   S. Dodge, P. Laube, and R. Weibel, "Movement similarity assessment using symbolic representation of trajectories", *Int. J. Geogr. Inf. Sci.,* vol. 26, pp. 1563-1588, 2012.
[http://dx.doi.org/10.1080/13658816.2011.630003]

SUBJECT INDEX